Why Is There Something Rather Than Nothing?

23 Questions from Great Philosophers

Leszek Kołakowski

TRANSLATED BY
AGNIESZKA KOŁAKOWSKA

BASIC
BOOKS

A Member of the Perseus Books Group
New York

Published in the United States in 2007 by Basic Books,
A Member of the Perseus Books Group

Copyright © Leszek Kolakowski, 2004, 2005, 2006, 2007
Translation copyright © Agnieska Kolakowska, 2007

First published in three volumes in Poland as
O co nas pytaja wielcy filozofowie, 2004, 2005, 2006
This translation first published in 2007
Published by Allen Lane, an imprint of the Penguin
Group in England in 2007
British ISBN: 978-0-713-99925-9

Books published by Basic Books are available at special
discounts for bulk purchases in the United States by
corporations, institutions, and other organizations. For more
information, please contact the Special Markets Department at
the Perseus Books Group, 2300 Chestnut Street, Suite 200,
Philadelphia, PA 19103, or call (800) 255-1514, or e-mail
special.markets@perseusbooks.com.

A CIP catalogue record for this book
is available from the Library of Congress.
ISBN-13: 978-0-465-00499-7
ISBN-10: 0-465-00499-7

10 9 8 7 6 5 4 3 2 1

Contents

Contents

Contents

Introduction:
What These Essays Are For

In introducing these brief essays about great philosophers – men who opened up new directions of
thought for future generations – I should start with
a *caveat*: my aim is not to provide a history of
philosophy in a pill. This little book is not meant
as some sort of super-condensed textbook, encyclo-
paedia or dictionary. If a student attempted to sit
an exam on the basis of these essays, he would be
disappointed: he would fail. There are plenty of
good textbooks, encyclopaedias and dictionaries of
philosophy, and I do not intend to 'summarize'
Plato, Descartes or Husserl; that would be an absurd
ambition. I would like, rather, to approach these
great philosophers by concentrating on one idea in
the thought of each – an important idea, an idea
that was fundamental to his philosophical construction, but also one that we can still understand today;
an idea that touches a chord in us, rather than being
simply a bit of historical information. I will try
to end each essay with questions addressed to the

reader: questions that arise from the thought of that philosopher and are still important and still clearly unresolved.

The order of these essays is chronological, with one or two exceptions. But this is not important; I want to talk about the thought of great philosophers, not about their lives, their intellectual predecessors, their influence or their place in history (unless these things have some importance for our understanding of their thought). As to who should be considered a great philosopher and how we can decide – well, those who like to spar about these things will probably go on sparring, but I think there is agreement for the most part. So I stick to my choice – it is what it is, and I will not argue about it.

Note on the English edition

In the original Polish edition there are thirty essays. But publishers are cruel beasts, and they demanded a selection. The philosophers left out of this edition, for one reason or another (but through no fault of their own), are: Aristotle, Meister Eckhart, Nicolas of Cusa, Hobbes, Heidegger, Jaspers and Plotinus.

Truth and the Good:
Why do we do evil?

SOCRATES
469–399 BC

My first great philosopher is of course Socrates. The two great pillars of European culture, Jesus and Socrates, never wrote a word; we know them only through secondary sources. In Socrates' case these are, of course, chiefly Plato's dialogues, but also, in slighter measure, the writings of Xenophon. How much of Plato's Socrates is the real Socrates speaking, and how much Plato putting words into his mouth, is a question that has elicited endless debate, which continues to this day. But it is not a debate I am competent to engage in. I want to raise a matter that comes up in the early dialogues, considered to be the most reliable in recording the thought of the true Socrates.

Socrates is not the first great mind whose ideas we are familiar with, but he was perhaps the greatest

architect of European culture, and is regarded as such even by those who do not share his philosophical views. If he is the great European master, it is not so much because of some specific doctrine he expounded as because of his *way* of seeking the truth.

What was the truth he sought? As a young man he was interested in the study of nature, but he later abandoned this and decided that the proper object of study was not physical reality, where everything changes and ultimately dies, but rather that which is unchanging – the immutable. This meant ideas. But not just any ideas; it meant certain basic ideas, most of them to do with morality. Socrates wanted to know what justice is, and virtue, and courage, and equality. However, he was not concerned with the meaning of these words as they are used in everyday speech; he wanted to know what such things as justice or equality *really* are – what they are in themselves. He thought that our understanding of such ideas is prior to what we know from observation: if we can say that two pieces of wood are of equal length, it is because we *already know* what equality is. In his insistence on the need to examine the meaning of fundamental concepts, and his belief that this meaning derived not from linguistic convention but from reality, the reality to which words refer, Socrates anticipated Plato's

theory of forms, or ideas – although, as Aristotle observes in the *Metaphysics*, he did not think that justice or courage were entities with an independent existence.

How do we go about discovering such truths? By endless, insistent, relentless questioning. In his role of street sage, Socrates grilled his interlocutors remorselessly, probing them, forcing them to go on asking questions that went ever further, ever deeper. His thought moved and took shape in constant dialogue – a dialogue in which he mostly asked the questions, while his interlocutors tried to answer, often saying only 'yes, Socrates', or 'of course, Socrates' or 'naturally'. Often in the course of the dialogue a new question would emerge, but without resolving anything.

It sometimes seems as if Socrates is only pretending not to know the answer (it was he who used to say, famously, 'I know that I know nothing') in order to force his interlocutor to engage in dialogue and reach a truth by himself, or revise his false opinions. He wanted to be a midwife like his mother: to coax truths out of their hiding place, where they lie ready to emerge into the light of day, waiting for us to grasp them and bring them out. He did not aspire to be original (but nor did any great philosopher: with such an aspiration he would not be ranked among the great); he wanted

3

only to grasp the truth and learn how to serve the good – for the truths that mattered were always truths about the good life. He lived an ascetic life, in accordance with his teaching: he taught that we should not care about worldly goods or the pleasures of the flesh, and he himself did not care about them. Unlike the Sophists, he took no money for his teaching; he lived in poverty, without complaint. And he showed courage when confronting his opponents, whether in war or in verbal sparring. The Delphic oracle decreed that there was no man wiser than Socrates – a fact he did not neglect to mention in his speech before the court, although he must have known that these words would be damning in the eyes of the judges and would only worsen his case, like his assertion that God had sent him to Athens to buzz around like a gadfly that spurs on a lazy horse by its bite. He also mentioned the voice of his 'daimon', an inner voice he followed that kept him from evil. We do not know what this daimon was, but presumably it was not Socrates' own creation (otherwise why should it have any authority?) but a moral force of divine origin.

But if Socrates thought he could lead his interlocutors to the truth by his questioning, it was because he believed that the truth is already in us, albeit often unknown to us: it accompanies us as we move on from our previous incarnation, so that

when we learn the truth we are not really learning it, but remembering what we once knew and have forgotten. And it is with the help of reason that we can learn all that it is important for us to know, and distinguish between good and evil. For what we call good and evil is not a matter of convention or agreement, nor even the result of divine decree. The sacred, Socrates said, is not sacred because the gods love it, but the other way around: the gods love what is sacred just because it *is* sacred. The sacred or good is that which is sacred or good in itself, regardless of what the gods decree and regardless of what we may imagine or agree on among ourselves. (This is a theme that continued into modern philosophy: Leibniz agreed with Socrates, and so did Kant.) Knowing how to distinguish good from evil is crucial in life. But it is no less crucial in the afterlife, for our souls remain intact after our death; they are not dispersed to the elements, even if we should die in a great gale. Good people, when they die, are happy in the company of gods; but evil people are unhappy and must wander for a long time as ghosts, aimless spirits loitering among the graves, until at last they are reincarnated, but as lower beings.

Because Socrates loved reason (though he may not have defined it clearly), and because he believed that truth, any truth, can be attained only through

a sustained effort of reason, the charges brought against him were perhaps not entirely unjustified. He was accused of failing to respect Athens' gods and instead worshipping others of his own invention, and of corrupting the youth with his talk. These accusations may not have been entirely accurate in the literal sense, but it was true that Socrates did seek truth through reason, not in tradition; he did encourage people to question everything; and he did try to inculcate this spirit of questioning in young people. Everything, for him, was at the mercy of reason; he recognized no other authority. But he did want to respect the law; indeed, he respected it even to the extent of refusing to escape from prison as his students urged him to. He knew he was unjustly accused, but believed he should obey the law even at the cost of his life.

It seems reasonable to suppose that someone who thinks that tradition should be worshipped for its own sake, and its alleged wisdom unquestioningly accepted as such – just because tradition has passed it down – will also think that the Athenian court was right to condemn Socrates to death.

It was because Socrates worshipped reason, because he identified reason with virtue, and virtue with happiness and human flourishing – genuine flourishing, which animates the soul – that Nietzsche hated him. Nietzsche claimed that truly

great spirits, destiny's elect, are guided by instinct, and that Socrates' cold, self-conscious reason, hostile to instinct, was a sign of decadence. He thought Socrates dragged the Greek spirit down into decline; he saw in him – and in Plato – a symptom of the decline of the old aristocratic culture. He even entertained the possibility that Socrates may not have been Greek; after all, we know that he was ugly, and ugliness, according to Nietzsche, was a symptom of the degeneracy that is typical of hybrids: 'monstrous face, monstrous soul', as the ancient proverb had it. Thus the best of men, punished by death for his love of reason, was accused of being a source of spiritual decline.

Here is a question that arises from Socrates' cult of reason:

Socrates maintains that it is not possible voluntarily to do evil if we know it is evil. If we do evil, it is through ignorance; if we know what the good is, we will do good. This may seem implausible to us: we tend to think that when we do evil, it is often when we are in the grip of a passion – hate, love, envy, greed, lust, desire or lust for power; that we know what is good but succumb to our passions nevertheless, unable to resist. We are inclined to agree with Ovid: 'I see and praise what is better, but follow what is worse.' Or with St Paul in the Epistle to the Romans: 'For I know that in me (that

7

is, in my flesh) dwelleth no good thing: for to will is present with me; but how to perform that which is good I find not. For the good that I would I do not: but the evil which I would not, that I do' (7:18–19). This seems common sense to us. But let us pause and consider for a moment whether Socrates might not be right after all. Perhaps the evil we do does have its roots in ignorance; perhaps our feeble reason is simply incapable of distinguishing between good and evil? If this were the case, would it follow that we are always blameless, whatever we do?

Being and Non-Being:
What is real?

We now go back two generations to Parmenides of Elea. I confess I would rather not mention him at all, for so many scholars, much scholarlier than I, have struggled to penetrate his thought and argued about the meaning of the paltry few dozen hexameters that survive of the poem in which he set out his metaphysics. But he cannot be avoided, for it was the power of his ideas that gave momentum to the great current of European thought, which went on flowing not only through the thousand years that elapsed from his time up to the end of the Athenian Academy, but for centuries after that, in Christian (but not only Christian) philosophy and theology, up to our own day. This current was a current of reflection about the One, the unity

9

that is hidden behind the diversity of our world of experience.

Ancient texts contain a wealth of allusions to and quotations from Parmenides' obscure metaphysics. We also have Plato's dialogue *Parmenides*, where Parmenides, by now an elderly gentleman, discusses the question of the One and the many, of the general and the singular, with his pupil Zeno, a generation younger, and a 'very young', perhaps twenty-year-old, Socrates. From the purely chronological point of view such a meeting and such a discussion, although possible, are unlikely; historians doubt that either ever took place. Plato's dialogue is deeply disturbing and entangles the reader in so many winding trails and snares that it is almost impossible to summarize it, but thankfully there is no need to do so here.

Parmenides' poem describes how some maidens, daughters of the Sun, brought him before a goddess who greeted him kindly and explained the difference between truth and mortal opinion — which, she said, is not truth at all, but only what seems to be. Parmenides wants to distinguish between what is, truly is, and what only seems to be. He thinks that what is, is necessarily; it cannot not be. Parmenides has been accused of confusing the existential sense of the verb 'to be' (as in the phrase, 'God is', meaning 'God exists') with its predicative sense (as

in the phrase, 'honey is sweet'). Perhaps he did confuse them, but this does not matter for our understanding of his metaphysics. Parmenides just says 'is', without a subject – '[it] is or [it] is not' – but what he means is that 'to be' in the fundamental sense is 'to be' necessarily: to be unable not to be. What truly is, is necessarily; it could not not have been. Its being, the nature of that 'is', cannot be grasped by our senses: it is accessible only to reason. But reason cannot grasp not-being, or what is not. The things we grasp with our senses come in and out of being: one moment they are, and the next they disappear; and we cannot say that something both is and is not, for that would be a contradiction. But what truly is cannot have been created, for that would mean that something comes from nothing, and that is impossible. Nor can it change, decay or die: it is perfectly fulfilled and unchanging, with no beginning and no end. Nor can it be said to 'have been' or to be 'going to be'; it simply is, beyond time, without time. At one point Parmenides compares this Being, the One, to a sphere. But this seems an awkward analogy, for 'being' in this absolute sense is also beyond space; he probably means that Being is full and sufficient unto itself, a closed unity, always the same, in whatever direction we reflect upon it. Not only is this Being, immobile in its eternal identity, accessible to reason alone, not

to the senses; it seems also to be the *only* reality that reason has access to: Being and the thought that grasps it are yoked together in such a way that in a sense they are one and the same.

Although the objects of our senses, the world of experience, are only names, not what truly is, nevertheless Parmenides permits himself to make observations about cosmological matters. This, some commentators have remarked, is not quite in keeping with his metaphysics, from which reflections upon what only seems, but is not in the true sense, should be eliminated. But Parmenides' significance for European thought does not depend on these few rather obscure sentences about astronomy. What makes him important is his attempt – perhaps a desperate attempt – to express something that probably cannot be expressed, namely what it means to say that something 'is' rather than 'seems'; that it simply *is*, without restriction or qualification. Being can neither shrink nor increase; it is also inevitably, inexorably turned inwards on itself. But there is no hint in Parmenides that this Being might be God, for there is nothing personal about it; there is perfection, but it is a perfection that arises from an absolute, secure immutability. Like the poet and philosopher Xenophanes of Colophon, some fifty or sixty years older than Parmenides and possibly his teacher, Parmenides rejected the traditional

Greek mythology, the gods of Hesiod and Homer. It was Xenophanes who said, famously, that if oxen, horses and lions could draw, they would draw their gods in the form of oxen, horses and lions.

It is hard to grasp the process whereby our reason penetrates what is, but we can try to reconstruct it hypothetically. If the world of our experience does not really exist, if it is just an illusion, like the 'maya' of ancient Vedic philosophy – an illusion of multiplicity which hides true Being – then surely this world, the world of all our perceptions and imaginings, cannot at the same time be Everything – everything that is; for then we would be forced to embrace the absurd conclusion that nothing exists. Since such a conclusion is unacceptable, and indeed probably incomprehensible, we are compelled to admit that what really *is* is radically different from all that our eyes see, our ears hear, and our hands touch; that this true Being, which can be neither seen nor heard, and is not subject to death or change, nevertheless *is* in the most primary, the most real sense. And since we can neither see nor sense it, but are compelled to infer its presence, it must, clearly, make its presence known to our rational thought – our reason.

Granted this, we still do not know how our reason can make contact with this perfect unity of Being. But the phrase 'make contact with' is

probably unsuitable: reason simply comes to know it by the compelling force of logical movement. More importantly, however, we also know nothing about the relation between the One and multiplicity – the objects of our perceptions. Is it possible to grasp how true Being, immutable and closed in upon itself, could create an infinite multiplicity of changeable and impermanent things? How could it do this without going beyond itself? And without losing anything of its immutability? This question arises not just in response to Parmenides (who does not address it explicitly in the fragments of his writings that have survived); it crops up, with disturbing insistence, throughout Christian theology. We also cannot say that Being is what is beyond appearances, for the very word 'beyond' refers to a certain spatial relation, but in this case of course no spatial relation can obtain. Similarly, and for the same reason, we cannot say that appearances 'veil' or 'hide' Being from us. Any attempt to describe this situation must be metaphorical at best, and the metaphors we must resort to cannot be translated into non-metaphorical language (in the way that we can, for instance, translate the sentence, 'He shut himself up in an ivory tower' into non-metaphorical language); they must remain metaphors, indecipherable, and we can never be certain that we have understood them correctly. The very word

'appearances' also gives rise to doubt, for it seems to suggest that an appearance is an appearance *of* something – something that itself is not an appearance. Perhaps (and this is a subject on which Bergson had much to say) it is just that our language is forced to labour under the weight of our spatial imagination and finds it hard to break free of this habit. Scholars have debated whether this world of appearances is real on some level – even the very lowest, even just a relative one – but according to Parmenides, being is indivisible and cannot have levels of existence: something either is or is not, there is nothing in between. And Being is everything it can be.

Parmenides' metaphysics perhaps achieved clearer and more consistent form in the writings of his follower Zeno of Elea, who left to posterity arguments (transmitted mostly by Aristotle) intended to demonstrate that motion and multiplicity are impossible. These arguments were by no means intended as jokes or as logical exercises. Proofs of the impossibility of motion had been discussed for centuries, and they were connected to the problem of the infinite divisibility of space and time. The Cynic Diogenes of Sinope is alleged to have refuted them by taking a stroll. This was not, however, considered a crushing and definitive refutation.

But what (someone might object) is all this for?

What good is all this idle speculation about unfathomable Being and the unity that hides behind multiplicity? It won't (they might say) keep you warm or put food on the table. But this would be a very silly objection. Certainly, it won't keep you warm or put food on the table. But the same can be said of vast tracts of culture. Yet in spite of this people have gone on asking such questions for centuries, undeterred by the thought of their (real or imagined) purposelessness. Why was this? Were they all just very stupid? No; they wanted to put into words something that our minds invariably encounter whenever they can snatch a moment of leisure from considerations of dinner and warmth, when they – these minds of ours – are mysteriously drawn to the peculiar question of Everything. And the question of Everything – the question of what is real – is the very question we have been discussing; Parmenides' concern can surely without distortion be rephrased in this way. We could concoct some definition of reality that might, on the face of it, be coherent and might, on the face of it, assuage our curiosity, but an arbitrary definition is not what we want; we want to know why the question of 'what is really real' still fascinates and bothers us.

But we have to admit that, although it strikes some sort of intuitive chord in us, it is hard to reformulate in such a way as to avoid the risk of a

vicious circle. We can ask, for instance, whether our dreams or hallucinations are real. Someone might reply that these are events which really occur, so yes, they are real. Someone else might say that they are not, because they only occur for me, not in the collective consciousness. Are we to say, then, that only those things are real which are accessible to the collective consciousness? Or perhaps we should rather say that *all* such things are real, not *only* such things? Let us consider for a moment what consequences follow if we accept such a premise. Am I real? Are you real? According to what rules do some say that God is real while others deny it? Both have their rules, but how can we know if they are sound? How is it that this curious intuition, encountered in ancient religions and in philosophy – the intuition that the world of our experience is not real – could strike people as true? It does not strike you as true? You think it is patent nonsense? Devote an hour to thinking about it, just the same. This is the question that Parmenides and his goddess suggest to us.

Change, Conflict and Harmony: How does the cosmos work?

HERACLITUS OF EPHESUS
c. 540–480 BC

Heraclitus was sometimes called 'the weeping philosopher', possibly because he often lamented human stupidity. He was also called 'obscure', and indeed his style was, like the Sibylline oracle, murky, ambiguous and full of riddles. The fragments of his work cited by later writers tend to be aphorisms, some of them now popular and widely recognized, like the maxim 'everything flows' (or 'everything is in flux') or 'you cannot step into the same river twice'.

Sayings of this kind have contributed to Heraclitus' image as an exponent of eternal change and impermanence. The image fits, but only up to a point. He did talk of eternal change, but he never said, like his follower Cratylus, that we can never step into the same river twice because it is constantly

flowing around us; nor did he ever say that, since nothing is permanent, we should not speak but rather communicate by gestures, for knowledge is impossible. From the obvious fact that everything changes (though at varying rates) it by no means follows, according to Heraclitus, that the things we see are illusions; for although, as he said, 'eyes and ears are bad witnesses' (though eyes are better than ears), and although only God knows everything, nevertheless we too – or at least a few elect among us, not the common people – can have access to wisdom. And wisdom is understanding Logos. In Heraclitus' case, this word of many meanings is usually translated as 'measure' or 'order' – an arrangement of things thanks to which the world, in spite of its mutability, is an ordered unity. Three elements – earth, fire and the sea – mingle and mutate into each other; fire is the primary element and the most active reality, although it does not seem to be the universal substance of which all matter is constituted.

But constant change – the continuous shifting of qualities into other qualities – is not chaos; there is proportion and equilibrium in the changing. In everything we can see a play of opposites, without which the world would collapse. There are simple ones like the beginning and end of a circle, both of which can be at any point on the circle. Others

consist in the gradual change of something into something else, like the change from hot to cold. Others still produce tension, the kind we produce when we shoot an arrow from a bow. We also see the interplay of opposites in the fact that the same things operate differently in different creatures: seawater is good for fish but bad for men, and so on. We may suppose that another of Heraclitus' famous aphorisms, 'the road uphill and down is one and the same', is also about how opposites mutate into one another. Heraclitus expresses this most clearly in his saying that war is the father and the king of everything; 'war' here seems to mean the conflict of opposite forces in nature.

Many of Heraclitus' maxims may seem like platitudes, some simply because they are so well known and instantly recognizable in their succinctness, others because they seem to be saying something utterly obvious. But Heraclitus' strength lies not just in his rhetorical talent and knack for the succinct phrase. He passed on to future generations a belief in the order of the world, order that can be hidden or partly hidden (for nature has a tendency to hide, and an invisible harmony is stronger than a visible one) and to whose force everything is subject. The phrase 'law of nature' does not yet appear, but Heraclitus talks of the 'Cosmos' – a universe that is ordered, but exists in constant

motion and conflicts of opposites. This order was not created by any divine or human hand; it always was and always will be. We can see that Heraclitus' way of thinking about the world is the opposite of Parmenides': multiplicity and conflict are what is real; tension and motion are real, and our knowledge of them is real too. There is no mysterious unity to be sought, immobile, incomprehensible and alien to life; motion and change are not veils of illusion. There are some who think that Plato's Theory of Forms – immutable entities which are contrasted with the multiplicity of impermanent particular things – was elaborated as a response to the Heraclitean world; it is a response that is different from Parmenides', but it, too, is the response of a seeker after the immutable.

Is God immutable, or gods? Heraclitus sometimes uses the word 'god' in the singular, but it would be rash to attribute to him belief in a single God who would be recognizably the same as the divine being of the monotheistic religions. God is night and day, winter and summer, war and peace, hunger and satiety. Is God, then, simply everything that exists, in all its multiplicity? But in that case the word 'God' would be misleading and unnecessary, for a belief in God's existence would add nothing to the universe we know. We lack the key that would unlock the riddle of how Heraclitus' 'god'

relates to Cosmos and to Logos. We know that Heraclitus had only contempt for popular religiosity and superstition, like praying to images; and he did not really believe in the 'truth' that people thought they could obtain from the Orphic or Dionysian mysteries. But we may surely take it that the word 'god' is not there merely as a literary ornament.

It is also hard to tease out from Heraclitus' words any coherent theory of the human soul. We know that the soul is unfathomable and its limits immeasurable. We also know that the best and wisest kind of soul is a dry one. What a dry soul is we do not know, except that a drunk man makes his soul wet. But a dry soul is not simply the soul of a man who is sober; presumably it is the soul of someone who is in control of himself and his emotions, thinks rationally and is not swayed by passions. The soul is made of fire, and when fire changes into water, the soul is lost, or dies; but it seems safe to assume, by extrapolation, that after another such change it returns and is found again. The better souls do not die with the body; and souls that die in war are purer than those that die as a result of sickness.

Aristotle took a dim view of Heraclitus' philosophy because he thought it denied the law of non-contradiction – the basis of rational thought. But when Heraclitus said that the same thing could both be and not be, what he meant was surely not that

we may make logically contradictory statements, but only that in the eternal, necessary and Logos-directed process of change, different qualities may gradually be transformed into their opposites, and in that transformation something can appear to be both such and not-such at the same time.

Let us end by asking, in spite of the murkiness and ambiguity of this obscure philosopher, a question that, if we knew the answer, would explain the whole of his metaphysics and his cosmology: whether some divine design can be discerned behind these eternal changes, behind the all-consuming war of things in eternal conflict, a design that not only encompasses the Cosmos but also gives it participation in good and evil. Do such attributes have any place in the Heraclitean world?

We may look for a hypothetical answer in one sentence which perhaps more than any other encompasses this whole opaque world. It is a sentence quoted by the Neoplatonist Porphyrius and it says, 'For God everything is beautiful, good and just.' What can this mean except that, if the world is so for God, then it *really is* so: beautiful, good and just?

This, then, is the question that arises when we reflect upon the Heraclitean Cosmos: can we make the same statement about the Christian God? Can someone who has faith in God in the Christian

sense describe the divine view of the world in the
same way? God knows all the evil that is in His
human subjects, He knows the ugliness and injustice
of what they do; what would it mean, then, to say
that in His eyes everything is beautiful, good and
just? For these purposes we need not worry about
the theological conundrum about God and evil: if
God is everything and encompasses everything,
and knows everything without ever going beyond
Himself, how could He know evil and injustice
without thereby Himself becoming evil and unjust? Let us set this problem aside and content
ourselves with construing God's omnipotence in
the ordinary way. What would it mean to say that
in God's eyes – in other words, in reality – everything is beautiful, good and just? And what can it
mean when Heraclitus says it (for he gives no
explanation)?

And if we do not reject it as obscure nonsense,
does it make sense to attribute such a view of the
world to the God of Christianity? Could such an
idea, which flies in the face of common sense, have
a place in Christian belief? That is the question.

The Good and the Just:
What is the source of truth?

PLATO

c. 427–347 BC

The whole of European philosophy is a series of footnotes to Plato, said the philosopher Alfred Whitehead, who himself had the reputation of a Platonist. And indeed, whenever we reflect upon an important problem in metaphysics, in the theory of knowledge and cognition, in ethics, often also in political matters, and in dialectics (which is the art of learned conversation), and in rhetoric, if we reflect upon the history of that problem, we will surely be able to trace it back to Plato.

Not everything in Plato has remained vital for us. His dreadful utopia, the vision of an ideal state, is historically interesting but not intellectually stimulating, in the sense that we do not wonder whether such a state would be a good thing and whether we would like to live there, because we

can see at once that it is a nightmarish invention, and that it would be a form of totalitarian slavery. But other ideas of Plato's do not seem outdated; they do not strike one at once as either obviously wrong or manifestly right. The Theory of Forms in particular raises questions that have still not been satisfactorily answered in a way that would meet with general consent. When people speak of 'Platonism', they usually mean the Theory of Forms, or Ideas – the entities later called universals, which are still debated today. In Plato's writings the Theory of Forms plays an important role in problems concerned with knowledge and cognition, but also in ethics and political theory.

According to Plato, true knowledge, as opposed to mere opinion, is about things that are immutable. Sense-perception does not give us knowledge in this sense; it gives us only fleeting impressions about ephemeral facts and mutable, impermanent things. We can truly understand those facts and things only by reference to Ideal forms, which are autonomous entities existing beyond the realms of time and space, independent of the physical world and access-ible only to reason. If such entities did not exist, no knowledge would be possible. When we say that something is just, for instance, we can know this only because we refer particular facts, actions or arrangements to the concept of justice – to the

immutable idea of justice. Moreover, even physical objects, even man-made artefacts such as tables, chairs and beds, are what they are – tables, chairs and beds – because there exists an essence of tables and chairs, a kind of tablehood and chairhood, of which these mundane objects somehow mysteriously partake. The same may be said, with greater conviction, of the objects of mathematics and geometry. A circle, as an object of true knowledge, is not this figure I have just drawn on this bit of paper with the aid of a compass. The figure I have drawn is merely an imperfect representation of the ideal circle, of the essence of circlehood, and this ideal circle has none of the imperfections of my drawing; invisible and indestructible, it can be known by reason, although it exists independently of reason and independently of whether anyone thinks of it or knows it.

The seventh book of Plato's *Republic* contains the famous myth, or allegory, of the cave, which is supposed to show us the difference between knowledge and mere opinion. We are asked to imagine some people sitting in a deep cave, where the light of the sun does not reach. They are chained in such a way that they cannot turn their heads, and can see only the wall – the far end of the cave. Behind them a fire burns, and between them and the fire is a low wall and a path along which other people are

moving, carrying statues or carvings of animals. The chained prisoners can see only shadows flickering on the wall in front of them, and they think these shadows are the only reality. Even if one of them could free himself and leave the cave, he would be so confused by the sight of the things of which previously he had seen only shadows, that at first he would take those shadows to be more real than the reality now before him; if he were to step out into the sun, he would be blinded by its light, and it would be a long time before he was able to see the sun itself. And if he were then to go back down into the cave, his fellow prisoners would make fun of him and ridicule what he said.

The allegory of the cave is about us – prisoners of the world of appearances, still shackled by their chains. The sun into whose light only a few are admitted is the divine idea of the good, the eternal source of truth and illumination for the mind. It is the idea of the good that tells us what the good life is and how to coexist with other people. But those who have seen the idea of the good expose themselves to the mockery of those who have not, and who know only shadows. So it is the duty of lovers of truth and wisdom, namely philosophers, to see to it that people – not everyone, just the best – see the source of the good and learn about it, and come to understand what matters most: what is just,

what is good and what is evil. To this end they must familiarize themselves with the world of forms – that other reality, eternal and infinitely better than our imperfect, fleeting, ephemeral world of shadows.

Reflecting upon the world of forms is of great benefit, according to Plato, in the study of politics and the best form of rule for the state. Those who love truth and know the good do not want power and do not seek it, yet it is they who would make the best rulers and guardians of peace, ensuring that the state was free from internal conflict; while those who have not attained knowledge, those whose reason is poor and who are consumed with lust for power and riches, will divide the state if they attain power, causing upheaval and internal wars which in the end will destroy the state itself, and the lovers of power with it. The conclusion of this analysis is somewhat startling: it is the philosopher who should be king of the state, for it is he, the lover of truth, who knows what is truly beneficial and what is truly good for man; and it is he, as one who has found the way to the divine, to the eternal good, who knows how to order public life: in accordance with his knowledge of true justice and true happiness. There will be no salvation for the world until it is ruled by philosophers.

This conclusion seems startling not only because

it is hard to find historical examples which might confirm the wisdom of such a proposal (Marcus Aurelius was a philosopher king, but there have been few apart from him), but also because in order to rule a state it is not enough just to have metaphysical knowledge, to know the eternal idea of Truth and Beauty; one must also have a lot of other specific and practical talents and abilities which metaphysics cannot provide. It is probably true that those who are driven only by lust for power and riches would, if they achieved their goal and attained power, soon bring about the downfall of society as well as their own; but it does not follow from this that those who care nothing for power or riches would make splendid rulers if by some miracle (for only a miracle would bring it about) they could attain power. The intellectual aristocracy is far from being suitable by nature for controlling affairs of state.

Plato himself was well aware of the objections against his Theory of Forms, and one can track the changes in his position through his work· in the *Timaeus* his approach to what can be known through experience is different from what it had been in earlier dialogues, while in the *Parmenides* we find a number of critical remarks – some of them devastating   about the Theory of Forms. Nevertheless, certain crucial ideas which formed

the basis of that theory were bequeathed to future generations and passed down to our own day. The so-called controversy about universals, a controversy which perhaps more than any other shaped the intellectual climate of the Middle Ages, has not withered away, though it is now somewhat differently expressed.

Certainly, we don't need to believe in some kind of ideal model of a table in order to know what a table is; and it is hard to see how such a universal – the idea of a table, or tablehood – could be useful in our thinking. We are not even sure quite how such a bizarre entity should be understood: it would be a table possessing only those properties which are common to all tables, but none which might distinguish one table from others, so it could not be square or round or rectangular, or big or small, or metal or wood, etc. What kind of table, then, would it be? It is also hard to agree that eternal forms not only exist but possess the attribute of which they are the idea: that an eternal form of beauty, for instance, not only exists but is itself beautiful. If we say this, we are at risk of falling into an infinite regress, for we are saying that something is beautiful and referring that something back to itself: the idea of beauty.

Nevertheless, two of the problems which emerge from the controversy about ideal forms are still vital

and can still disturb our minds' slumbers. They encompass two vast tracts of culture which seem to depend on a belief in timeless, ideal models, and the question we must answer is whether we can do without a belief in such models.

The first concerns the ideas of good and evil. Thrasymachus, one of Socrates' interlocutors, thinks that justice is whatever is in the interests of the stronger, and Socrates argues against this. But we are not concerned with his arguments here. What concerns us is this: if justice is whatever the stronger – a tyrant, for example – declares to be just, then, since clearly tyrants will differ in what they consider just, there is no common idea of 'justice'; there are only conflicting interests and conflicts between despots, nothing else. So we can safely eliminate the word 'justice' from our dictionaries: it is a misleading word, which we use only to defend someone who is, or would like to be, a tyrant. It is not only Socrates who would be reluctant to embrace such a conclusion; many other people, today and in the past, would be equally reluctant. And many people would like to know what justice really is; what is really just and what is not. But in order to know this, they must believe that there is some idea of justice that has not been established by someone's arbitrary decree – an idea given in the world, as it were; they must believe that there is a

meaning to existence, a meaning bestowed by God, or by gods, or by some moral constitution, upon whose verdicts we may rely. The question, then, is this: is this the choice we are confronted with? Either that justice is an arbitrary decree which can be established by anyone with enough power, or that it can be known through some source of wisdom which lies beyond the world? Either that nothing is really just or that there must be some rule, some measure of justice, which was not invented by us and is not arbitrary, but Platonic? Or is there a third possibility?

The second question is about mathematical objects. Let us take some elementary mathematical truth, for instance the statement that seven is a prime number; and let us ask what this statement is about. What is seven, and what is a prime number? We can espouse a radically anti-Platonic view and hold that the only things that really exist are particular objects: that although we do, of course, use general terms when we describe these objects – since objects share many qualities – these are only names; nothing general, or universal, really exists. But what, in that case, is the world of numbers and other mathematical objects? In what sense, if any, can they be said to be real? Many philosophers have taken the view that we cannot avoid recognizing the reality of general entities:

Leszek Kołakowski

Bertrand Russell, one of the founders of modern analytical philosophy, was among them. He denied that universals were mental constructs: he tried to show, for instance, that similarity itself – not only similar things – is real. Some mathematicians and physicists were, or are – for there are many such today – Platonists with regard to the world of mathematical objects: the great Polish mathematician Jan Lukasiewicz was one example; Roger Penrose is another. We might also consider whether it makes sense to say that seven would still be a prime number if there were no people or any creatures endowed with reason, or even if the world did not exist? This, then, is the second question about Plato, related to the first: can we survive intellectually if we believe that the world contains nothing except individual objects, and in particular that there is no such thing as an invisible – albeit accessible to reason – kingdom of mathematical entities?

Life in Accordance with Nature: Can it make us happy?

EPICTETUS OF HIERAPOLIS
c. AD 50—130

Epictetus of Hierapolis, a Stoic philosopher and moralist, wrote about how to live. He wanted to teach us how to lead a life that is fully free, a life of dignity and unclouded happiness. Questions of logic crop up in his writings only insofar as they are necessary for the purposes of his moral arguments. He also says little about metaphysical questions, but what he says about them is important, for they show a continuity between Epictetus and the ancient Stoic tradition, particularly Zeno of Citium, the founder of Stoic philosophy, and Chrysippus. The ancient Stoics (admired by all logicians, for they were the first to construct a non-syllogistic logic) were said to have written vast amounts: we know of over 700 titles of Chrysippus' works, mostly about logic and language, but none of them

has survived; we have only a few fragments or second-hand summaries. Epictetus, on the other hand, wrote nothing (like Socrates, whom he worshipped); we know of his teachings thanks to the industry of an eager pupil who made notes of his lectures – probably quite faithful notes, for they contain many repetitions, which show that the text was not edited – and collected them in several books. The most famous of these is a small book called *The Manual*, a brief summary of Epictetus' moral teaching; a longer work, the *Discourses or Diatribes*, consists of a long series of more detailed treatises in dialogue form.

The chief principles of Stoic metaphysics (I disregard their speculations about physics and astronomy), which Epictetus clearly espoused, were as follows: the world is one substance, and its fate is ruled, in every particular, by Providence, which means God, which means Reason, which means Destiny, which means Order. Nothing in the life of the world is due purely to chance, nothing is inexplicable; everything has a reason, for everything is well guided and wisely arranged by the highest power. The world and our existence are not guided by blind causality; everything is part of a providential order, everything has a meaning and an end. Why, then, is there evil in the world? And why is there so much of it? Either because the perfection

36

of the whole for some reason requires imperfection in its parts, in the details; or because God, who always wants only good, must, in order to do good, use tools which cause suffering; or, finally, because we are endowed with freedom of choice. But our freedom is not in conflict with Destiny. The cosmos is subject to cyclical regeneration: when it reaches the end of a cycle, it is consumed in a great fire, whereupon it is born anew and history is repeated from the beginning, just as it happened, in every detail; there will be another (or rather, the same) Socrates and the same judges who will condemn him to death.

Upon this metaphysics Epictetus constructs his moral doctrine – a doctrine that is dazzling in its implacable consistency. Everything, he indefatigably repeats, falls into one of two categories of things: those that depend on us and those that are independent of us. Our thoughts, our judgements and desires, our acts of will and spirit, belong in the former category; our bodies, our health, fame, wealth, honours, worldly goods – the whole of our material existence – belong in the latter, for all these things can be snatched from us by fate. They are all subject to Destiny, over which we have no control. If we cling to these independent things and worry about them, our lives will be beset by perpetual anxiety; we will curse the gods, the world

and the people in it; we will be slaves to those supposed goods, which are not real goods at all. But once we understand this crucial distinction, we will attain a state of imperturbable indifference towards the things over which we have no control. No evil, no injury, can touch us then; we will be happy whatever happens. Ruin, bankruptcy, sickness or death will not affect us; neither the death of someone close to us, of a wife or child, nor our own illness and the knowledge that we will soon die, will be a source of sorrow. I can be tortured and sentenced to death, suffer hunger, abuse, slander or ridicule, yet I will not succumb to sadness or despair; nothing can disturb my serenity. For it is not such things in themselves that give rise to resentment and anxiety, but only my beliefs – my false opinions – about them. Once I cast off these beliefs I will be free, safe and content, whatever blows of fate befall me.

Since I know that Destiny controls everything, I do not expect things to conform to my desires or whims, on the contrary, I want everything to be just as it is, and am grateful to Providence for her gifts, even when those gifts come in the form of suffering, misery and death. I shall be like an actor on the stage, who must perform as well as he can a role written and allotted to him by someone else. It can be the role of a beggar or a king, it can be short

or long – it doesn't matter; I will be content if I play it well. In other words, I must accept fate and praise it, whatever it brings, for I know that it is determined by the gods, in their goodness and wisdom. I will always be serene, although I will refrain from smugness in my superiority over others. This is called a life lived in accordance with nature.

Such an attitude to the world allows us to achieve a kind of connection to God; we realize that we are part of the divinely decreed destiny of the universe. Thanks to this knowledge I am no longer merely a citizen of Athens or Corinth, but a citizen of the world; and I know that everyone is a child of God. My indifference to external things, to all those matters over which I have no control, does not estrange me from other people – participants with me in the same fate, the same universe. On the contrary, once I have grasped this unity of fate, I will be helpful and friendly towards others, even though I know that most of them are incapable of acquiring that ability which is proper to the philosopher – to affirm destiny.

This one phrase contains the core of the Stoic attitude towards the world, and encapsulates the whole of Epictetus' philosophy: the affirmation of Destiny and the love of fate – *amor fati*. This affirmation must be total if it is to be genuine, and

of course it can be genuine only if we believe that fate is not the result of atoms haphazardly interacting with other atoms, but the implementation of a divine plan. This plan is good by its very nature; evil comes not from the world but from my beliefs about the world. And it is entirely in accordance with this plan that I, too, should suffer misfortunes – troubles, sickness, death – for these are somehow instrumental in working towards the good of the universe. If we could know the future, we would ourselves make sure that all these misfortunes befell us, concerning ourselves only with those things that are truly worthy of concern – our souls and virtue; we would not rebel absurdly against fate, but would want it to bring exactly what it does bring.

The Stoics' moral doctrine, Epictetus' in particular, suggests a number of questions. The first is traditional and not specific to this doctrine; it arises from a variety of doctrines and has been debated for centuries by philosophers and theologians. It is this: if we assume that the world and everything that happens in it is determined in every detail by Providence, can we still, without contradiction, defend our belief in human freedom of choice and individual responsibility? If nothing except my opinions and judgements depends on me, in what sense can I be said to have free will? A traditional

question, I repeat, but one that arises ineluctably from the Stoic doctrine. It also arises within the framework of Christian theology, at least in some of its variants.

The second question, this time addressed to Epictetus in particular, is this: what, if anything, does it mean to say that since the past is gone and the future yet to come, death deprives us only of the present, and it makes not the slightest difference when and in what circumstances that present comes to an end?

And a third question, which goes to the root of Epictetus' moral doctrine. Assuming that we could achieve this superhuman indifference to all the things we can do nothing about – to the things that befall us, whatever they might be, to other people and to the world in general – would such imperturbability, such icy emotional impassivity, be desirable? And should it be called happiness? If Job had read Epictetus' doctrine (written several centuries later) and taken it to heart, would he have been happy in his torments? Could one be happy while suffering appalling tortures, being flayed alive or boiled in oil, in the knowledge that one's pain is part of a plan laid down by Providence? It is true that people have borne indescribable suffering in the belief that they were suffering in a good cause, but could they also rejoice in the very fact that Providence had

chosen to inflict those sufferings upon them? Is
there no difference between an impersonal cosmic
Reason before whose plans we must bow, though
we do not know what they are, and divine Providence as it is understood in the Christian faith – a
Providence whose designs we also do not know but
which we trust because we have faith in its goodness
and believe that it will lead us to salvation? (There
is no evidence that Epictetus thought of death as an
entry into a different world in which our individual
existence is preserved.) And can we, when confronted with the death of someone close to us, be
content in the knowledge that we must all die, so
there is no point in being sad and lamenting the
fact? Can we not only accept without demur all that
fate brings, but also derive happiness from that
acceptance?

To be sure, Stoic impassivity can to some extent
be recommended, and probably in some measure
achieved. Cultivating a Stoic indifference to twists
of fate and unexpected material losses, viewing
one's life as one would view a film, is probably a
good strategy for living; far better, at any rate, than
the feverish pursuit of so-called success, fame and
riches – a pursuit that in most cases is fruitless
and leaves only bitterness and a sense of injustice,
and ultimately of failure. But would such indifference be a good and desirable thing if it were total,

all-encompassing? Should we really aspire to remain untroubled and unmoved by death and suffering, even by the suffering of those close to us, just because these things are a necessary part of a cosmic design? Could we really find happiness in such indifference? And would it enrich our lives? Or would it, on the contrary, impoverish them and rob them of their fullness? Are ordinary human emotions – love and compassion, sadness and joy – simply proof of our ignorance and immaturity, rather than of life lived fully in the diversity of the real world? And why should the Stoic life be in accordance with nature, since it seems that nature made us quite differently, and the perfect Stoic is a rarity, a freak, indeed a monstrosity? These are some of the questions Epictetus raises.

Knowledge and Belief:
Can we know anything?

SEXTUS EMPIRICUS
2nd century AD

The popular sense in which we use the words
'sceptic' and 'scepticism' today is a loose one: when
we say that someone is sceptical about something,
a project, say, or an opinion, we are simply saying
that he has doubts about the truth or reliability
of that opinion or about that project's worthiness
or chances of success. But in antiquity, and also
in later centuries, a Sceptic was a follower of a
specific, well-defined doctrine about our cognition
and our knowledge – and also about the way to
live. A Sceptic was someone who examined things
carefully.

It is questionable whether Sextus Empiricus himself was a great philosopher; by his time the Sceptical School had existed for several centuries, but
the philosopher-sceptics before him either did not

trouble to write anything at all (like Pyrrho of Elis, the founder of Scepticism and the most famous and important of them – so important that people often used to speak of 'Pyrrhonism' rather than Scepticism) or little of it survived. Sextus, on the other hand, left behind him a considerable body of work, in which, with admirable industry and in elaborate detail, he sets out the sceptical doctrine. He is the only ancient codifier of that doctrine, and since a doctrine of such enormous significance as this one was to have cannot be ignored, its greatest exponent deserves a place among the great.

Sextus' writings contain hundreds of arguments the purpose of which is to lead us to acknowledge that everything we think we know is in fact uncertain; that the reasons for accepting opposing claims are equally strong (therefore none is really strong); and that different philosophical schools proclaim different, mutually contradictory, opinions. Some of these arguments are not confined to the Sceptics, but may be found in the work of various other philosophers: for instance, arguments about the fallibility of our sense perceptions. Animals also perceive through their senses, but differently from people; why should our perceptions be more 'true' than theirs? And people, too, differ in the way they perceive things, because of varying external circumstances or the context in which the thing is

perceived, but also because of sickness, age, and so on. We have no grounds for claiming that we really know things, that our senses penetrate to things as they really are; we know only appearances, only the stimulation of our senses. The Sceptic does not do away with appearances, he merely notes that they are as they are: honey appears sweet to us, but there are no grounds for accepting as true the claim that honey really is sweet in itself. A healthy person may see a certain surface as white, while someone who has jaundice may see it as yellow. If we say that a tower seen from a distance appears round, but polygonal when seen from close up, we are opposing two different appearances to each other, and neither of them is more justifiable or worthy of assent than the other. The same appearances can be pleasant for some people and horrible for others. But the dogmatists – who include other Sceptical schools as well as Epicureans and Aristotelians – make judgements about things that go beyond appearances.

Arguments for various metaphysical or theological claims are equally undecidable, Sextus says. Some people claim that gods exist, others that they do not; some people believe in Providence while others point to the misfortunes and evil that exist in the world as proof that Providence cannot exist, because if it did it would not allow such things. Nor

can general agreement be considered a criterion of the truth of an opinion, for one cannot know everyone's views on a given matter; and even if there were such consensus, it would not be a reliable criterion of truth. The distinction between dreams and our perceptions when we are awake also shows us how relative are our opinions about what is and is not real.

But there are other objections, even more fundamental than these, that undermine our claim to know anything. In order to be convinced of the truth of an opinion or belief, we must have signs with the help of which we can tell if something is true or false; in other words, we need a criterion of truth. But how can we tell if a given criterion is reliable? In order to be able to assess its reliability, we must have another criterion according to which we might judge it. And so on ad infinitum. Thus there is no criterion of truth; there are no signs that could tell us what is true and what is false. Similarly, in order to be convinced by a proof, we need proof that the proof is reliable. And so on ad infinitum.

Thus the opinions of philosophers about such things as the elements, or the soul, or the gods, are all worthless and without foundation; they are unjustified dogmatic assertions. This also applies to all our opinions about good and evil. Sextus invokes a wide variety of different customs and beliefs held

by different tribes or in different countries about what is allowed or forbidden in sexual life, or clothing, or food; their different laws, religious beliefs and rituals.

Is the Sceptic, then, compelled to declare that there is no truth and that knowledge is impossible? God forbid! When Carneades, one of Sextus' great predecessors, declared that knowledge is impossible, he demonstrated thereby that he was not a Sceptic. For the assertion that knowledge is impossible is no less dogmatic than Plato's or Aristotle's opinions about the hidden properties of the world, ideas and divine beings. Furthermore, such a claim entangles us in self-contradiction, for to assert that knowledge is impossible is to assert that one has acquired the knowledge that knowledge is impossible. Similarly with the statement that 'nothing can be known'. The Cyrenaic School claimed that all things in the world partake of some kind of hidden matter of which we can have no knowledge; the Sceptic does not make such dogmatic claims: he simply suspends judgement. The suspension of judgement is the Sceptic's principal intellectual act. His slogan is 'not more': for any given opinion, the arguments for it will have no more weight than the arguments against it. There is no need for the Sceptic to make general pronouncements about the value or worthlessness of cognition or the power of doubt.

Indeed, Scepticism is strictly speaking not a theory or a doctrine, but a method of doing away with all theories. It is, as Sextus gracefully puts it, a medicine that flushes harmful humours out of our organism and flushes itself out with them. Scepticism is a state of mental health in which medicine, including the medicine of Scepticism, is no longer needed.

But can we live well after such a cure, in a state of permanent suspension of belief? Certainly, the Sceptic replies; and not only that: it is the best kind of life we can live. The Sceptic does not say, like some philosophers, that our aim in life is pleasure. Our aim is quietude, or serenity – a serenity achieved through the sceptical therapy. And what could be better for the soul than serenity? But what are the practical rules which should guide the Sceptic in his everyday life, given that he knows that there is no truth (with the proviso that he himself does not make such dogmatic claims, only suggests them) either in the moral sphere or in matters of logic, or in the evidence of our senses, about how things really are? The Sceptic obeys the laws, customs and traditions of the city or country in which he lives. He is also subject to natural emotions. He lives, he says, 'according to appearances'. In order to lead a normal life we have no need, he says, of opinions about the existence or non-existence of the things around us; the Sceptic

uses those things or avoids them like everyone else, but he makes no dogmatic assumptions about their ontological status. Gorgias claimed that nothing exists; Heraclitus claimed that everything exists. But the Sceptic does not worry about such things, just as he does not worry about whether God exists, what His attributes are and whether He knows the future.

We must also forget about the things the logicians teach us. We have no reliable way of establishing definitions, divisions into categories, species and kinds. Sextus was possibly the first to notice something often remarked upon by modern philosophers, among them J. S. Mill, namely that even the simplest syllogism – all men are mortal, Socrates is a man, therefore Socrates is mortal – involves a vicious circle, since in order to know that all men are mortal, we also need to know that Socrates is mortal. Even mathematics does not escape the Sceptic's criticism. Nor do opinions concerning time, space and causality – questions such as whether time is real, whether it is infinite or infinitely divisible, whether we have knowledge of any causal relations.

Not all Sextus' arguments – and there are myriads of them – have survived down to modern times, at least not in the form in which he expounded them.

But many did survive, remaining vital and nourishing philosophical thought for centuries. What, then, are the questions that Sextus' arguments suggest?

First of all, we can ask whether the following is a sound and convincing argument: there is no criterion of truth, because in order to establish such a criterion, we would have no have another criterion, and so on ad infinitum? And if this is a good argument, what follows from it? It does not follow, according to Sextus, that reliable knowledge is impossible; such a claim would be dogmatic and self-contradictory (because it assumes that we can know that we know nothing and can never know anything). Is this last argument refutable?

It is important to remember, when considering these things, that the Sceptic makes no general claims of any kind; that his aim is to cure the soul and free it from the burden of needless and unresolvable worries, not to convince it of the truth of some doctrine. And this leads us to the second question. Sextus himself admits, though he does not elaborate on this point, that man is an animal which by nature loves truth. This seems right; it is probably true that we want to know what is and is not real, or true, not just because we need this knowledge for practical purposes but also for the sake of the truth itself. But is it possible for a lover of truth

to be, as Sextus would have him be, indifferent to the truth? Is it possible to achieve a mental state of indifference with regard to God, time, the soul, causality and thousands of other matters? To be sure, if I want to eat an apple, I do not need any knowledge about the nature of apples in themselves – as distinct from those properties of apples that I can know through sight or taste. If I want to build a stone house, I do not need to know about the nature of stone as such; the experience of cutting stone is enough for my purposes. But would such imperturbability, if it were attainable, really be the best, the most perfect kind of life? We might say: well, since we know nothing, what is the point of constructing theories that have no foundation? (Although here, again, a puzzle lurks: the Sceptic cannot say that we know nothing, he can only suggest it.) But if philosophers and scholars had seriously attempted to achieve such self-satisfied serenity, would they have been able to build our civilization? Would modern physics have been invented, with all its technical applications of which we avail ourselves every day, had it not been preceded by the speculative physics which Sextus accused of sterility and of making unfounded claims to truth? This is the second question Sextus suggests.

The third and last question is as follows: does

the Sceptic contradict himself in expounding the Sceptical doctrine? Should he not rather remain silent, if he is to be consistent?

God and Man:
What is evil?

ST AUGUSTINE
AD 354–430

Among the many striking statements in the writings of St Augustine that lodge in the mind – his memorable mental shortcuts, as it were – the most famous may be the sentence, 'I desire to know two things only: God and the soul. And nothing more? No, nothing at all.' That answer – 'nothing more', or 'nothing at all' – might suggest that, apart from those two principal aims towards which our thought strives, there are many other things we could think about. But no. To know God and the soul is to know everything: creation and its purpose, the order of the universe and the divine presence in it; our cognition and knowledge of the world, and therefore the idea of truth; the power of reason and the power of faith; man's place in the universe and the nature of time. Augustine's work is a monumental

edifice, an immense construction on which the intellectual life of Christianity – both theology and philosophy – reposed for centuries. His huge treatise *City of God*, the greatest work of Christian antiquity, came about owing to a cruel accident of history – the sacking of Rome by the Visigoths in 410. This accident gave rise to the Christian philosophy of history, and also to a powerful response to the problem – though it was not a new problem – of evil in the world.

The question of evil is ubiquitous in St Augustine's work. He himself, after all, before he converted to Christianity and was baptized at the age of thirty-three, the age of Christ, had been a follower of the doctrine of the Manichaeans, who believed, in accordance with Persian mythology, that good and evil are two independent forces engaged in an eternal struggle. But a Christian believes that there is one Creator, who is good and does no evil, although He allows it – either, as St Augustine writes, so that greater good may come from it, or in order that the contrast between good and evil might enrich the world. The very existence of evil in the world must be a good thing, since God, in His omnipotence, allows it. But evil cannot be a kind of being, for being has only one source; evil must therefore be pure negation, the absence of what should be: nothingness, void, an upsetting

of the order of things. Nature is not evil; it might
be corrupt and tainted by evil, but it is still good.
Nor is matter evil. Strictly speaking, evil cannot be
known, for one cannot know nothingness. But the
possibility of evil — in other words, of decay and
corruption — exists because everything except God,
and thus also man, was created from nothing, and
hence is subject to change.

Since God is absolute good, it follows that He
alone deserves worship, faith and love as ends in
themselves; we may love God's creatures, including
our fellow men, only by reference to Him. The
world is full of the traces of God; He is always
close, and constantly present in our souls, although
we can turn away from Him if we are guided not
by love but by desire, which is merely self-love.

We can see the traces of God in our souls
directly, because the soul is where truth resides.
This truth is not, as in Plato, a remembrance of
something once learnt — in a previous existence —
and later forgotten; it is the good with which we
are born, with which God created us. And because
we are capable of grasping eternal, immutable
truths, we are also able to grasp that we, too, are
immortal; for the soul, since it is the receptacle of
eternal truths, understands its own participation in
the eternal order. Our reason sees that its ability to
grasp eternal truths was not created by its own

powers but comes from that source which itself is the truth and the highest good. Another often-quoted aphorism of St Augustine's, from his treatise *On True Faith*, says: 'Do not go out into the world but return to yourself: the truth resides in man.' Indeed, if the truth resides in our souls, then God also resides there; and if we err, it is because we have sinned; because we have voluntarily turned away from God. Evil and falsehood are united in the same way as goodness and truth – which is equivalent to being and truth, for only good partakes of true being.

The arguments which the Sceptics use to persuade us that there is no truth are poor, miserable things. If I doubt, there is one thing I cannot doubt, namely that I doubt; I cannot, in other words, doubt the truth of my own existence. Here Augustine anticipates Descartes. The Sceptic refutes himself, he says, for in his very act of doubting he is grasping a truth within himself. Since we ourselves are receptacles of truth, we can, he says, free ourselves from the vain curiosity which draws us to ephemeral things instead of eternal ones.

The fact that we can discover a truth about God through our own faculties does not mean that we hold in our minds purely intellectual proof of the existence of God. Another well-known sentence of Augustine's, from his commentary to the Gospel of

St John, says: 'Do not try to understand in order to believe; believe in order to understand.' Faith precedes reason. But there is no faith without reason, for only reason can believe. This may seem a confusing intellectual construction, but the confusion is easily cleared up. After all, even the propositions of mathematics work like this: we believe them first, and only later do we try to prove them. God gave us the faculty of knowing the truth; He enlightens us. Therefore man, if he does not want to renounce God, knows what the truth is before he is able to assimilate that truth intellectually. The truths we are talking about here are all intellectual truths; we are not talking about the truths of the senses, the truths that come from experience: these do not come through divine illumination. But nor do our sense perceptions come about because physical objects cause changes in our souls; material things cannot causally affect the soul. The human body is the instrument of the soul, and when it experiences the influence of external bodies, the soul turns towards what is happening in its body, and in this way comes to know the material world. Thus the soul is always active in our acts of cognition.

The soul, unlike God, exists in time. It is everlasting, but not eternal in the sense in which God is eternal. God is eternal in the sense that He is timeless; for Him there is no future and no past. He

knows all of reality at once, in His eternal immobility; He encompasses everything, all that is past and future for us, in a single act of perfect knowledge, without going beyond Himself. This is something we cannot fully understand. Living in time as we do, we know that what is past no longer exists and what is future is still to come; human existence is locked into the present, as it were, enclosed within that infinitely thin line that divides the past from the future. But what is past, although it has no real existence, is nevertheless preserved through memory, which the human soul carries within itself. Similarly, the soul endows future events with a kind of existence through anticipation. St Augustine does not explicitly say that the continuing existence of everything depends on our consciousness, but he does suggest this. One might even suspect that time is a psychological reality, and that whatever is real is real only by virtue of our perception. But this is not the case, for we know that God created the world together with time, thus He also created time; therefore time must have its own existence, although of course that existence, like everything else, is dependent on God. At this point we come up against a mystery that we cannot fathom.

Since all that we know, we know thanks to divine illumination, there is no reason for Augustine to distinguish between philosophical and theological

knowledge; all that we know and all that we do will, one way or another, lead us to God. This does not, of course, mean that we cannot turn away from Him, that evil is not rooted in us. Not only is it rooted in us, but since the Fall it is omnipotent in us unless divine Grace stifles it. Everything we do of our own volition is evil, against God; whatever we do that is good is the result of grace. Grace by definition cannot be deserved; it is a divine gift, freely given. Moreover, it is irresistible: we cannot refuse it. This is a horrifying view of the world: if it were not for divine Grace, we would all be (deservedly) in hell; but God chooses some, a few, and grants them (undeservedly) a share in eternal happiness. His choice in no way depends on anything we might do; it is His free decision, the reasons for which are unfathomable.

Nevertheless, Augustine insists that we do have free will. For when God bestows upon us the gratuitous gift of divine Grace, we not only do good, but also will good; we do good voluntarily. Without Grace we do evil, and we also will evil; we do evil voluntarily. Left to ourselves, we have nothing except sin and falsehood. We fall by ourselves, through our own will, but we cannot pull ourselves up without God.

Since God lifts up some and leads them to salvation while abandoning others to evil and dam-

nation, mankind is divided in two: the kingdom of earth – all those, including the dead and the still unborn, from the day of Creation until the end of the world, who live enslaved by their own lusts and desires, who love the ephemeral and will ineluctably fall into the clutches of Satan; and the kingdom of heaven – all those, past and future, whom God has, from the beginning of time, intended for salvation. Until the Last Judgement, both will continue to live together on earth, mixing and participating in the same earthly concerns; but for the first group earthly goods are an end in themselves, while for the second they are only a means on the way to the kingdom of heaven. Cain belonged to the first group, Abel to the second.

For Augustine there is only one historical process; here, contrary to the doctrines of some Greek philosophers, there are no cycles, regenerations or returns. And the things we consider to be the results of chance are all parts of the wise plan of Providence, which is veiled from us.

Unlike some earlier Church Fathers, but also some medieval theologians and Reformation thinkers, Augustine not only devoted much attention to pagan philosophy but also took pains to demonstrate how close some pagans came, without knowing it, to Christian truth. Above all he admired Plotinus, whose doctrine he interprets – with some

exaggeration – in such a spirit as to make it accord (only in part, of course, for Plotinus did not know the Redeemer) with Christianity.

The questions suggested by St Augustine are endless. Here are just a few:

If we accept that everything we do of our own volition is evil and that we do good only thanks to the operation of irresistible divine Grace, can we still maintain, without contradiction, that we have free will? And if we believe that God is the absolute sovereign of all existence, that He chooses from among His subjects those who will enter the kingdom of heaven and abandons others to their own corruption, and that His choice has no connection with the deserts of either – if we believe this, can we also believe, without contradiction, in divine justice?

What is and should be, exists; what is but should not be, has no existence. That is what Augustine seems to be saying when he talks of evil as pure negation. So both good and good things exist, and corrupt things also exist; but corruption itself does not exist. Can we make any sense of this?

God's Necessity:
Could God not exist?

ST ANSELM
AD 1033–1109

Anyone who has ever had the slightest brush with philosophy will know, even without an intimate acquaintance with its concerns in the Middle Ages, that in the eleventh century St Anselm of Canterbury (or Anselm of Aosta, in Piedmont, for that is where he was born) constructed an extraordinarily interesting line of argument known as the Ontological Argument for the existence of God. The name was thought up by Kant seven centuries later, and since then that is how Anselm's argument has generally been known. It is a troubling argument, as we can see when we consider how many of the most resplendent stars in the European cultural firmament went to enormous pains trying either to refute it or to show, by amending it in various ways, that it was sound. Among them were philosophers of the

63

most varied orientations: Descartes and Leibniz, Kant and Hegel, Schopenhauer and Bertrand Russell and G. E. Moore; and one of the twentieth century's most distinguished mathematicians, Kurt Gödel, analysed the logical structure of Anselm's argument.

The argument itself, though a thick layer of commentary has grown up around it, is not difficult to understand. Briefly summarized, it goes as follows. Let us imagine a being such that no greater being can be conceived. Anyone hearing this description, Anselm says, will understand it; one needn't be particularly clever to grasp it. Now it is not hard to demonstrate that a being so conceived – and we can all conceive of it – cannot just be conceived, but must actually exist. For if it is conceived, and thus exists in our minds, we can imagine that it really exists; but a being that really exists is greater than one that is merely conceived by us and exists only in our minds. In other words, a being than which nothing greater can be conceived would, if it did not really exist, be a being than which something greater *could* be conceived (namely something really existing); it would therefore be self-contradictory, an impossibility. In order to avoid self-contradiction, we are compelled to admit that such a being must exist: that it exists necessarily.

God is just such a being – or rather, He is that being. Everything else we can think of can without contradiction be thought of as not existing, but not God. His existence is unfathomable. The fool who, as the Psalmist notes, 'hath said in his heart, There is no God', could not really have understood what it was that he had said in his heart; if he had been able to understand it, he would have seen that it was self-contradictory. So no experience is necessary to attain intellectual certainty of God's existence; we know it *a priori*, from the very idea of God.

We must not imagine, however, that Anselm needed this argument in order to be certain of God's existence. The context in which the argument is embedded is not intellectual exercise; it is an exalted state of prayer and contemplation, a plea by Adam's sons, condemned to a life of wretchedness and misery, to be allowed to approach the world of the divine. Anselm was a man of deep faith; he had no doubts about the existence of God, his certainty was unshakeable. But his faith was – in accordance with the original title of the treatise in which the Ontological Argument appears – a 'Faith in Search of Understanding'.

Anselm was one of the subtlest and most pen-etrating European minds of his time. Reason and reasoning were not the source of his faith; for him, faith, precisely because it is faith, needed no such

support. Nothing was more alien to him than 'rational religion', as it was later called – the approach, characteristic of deism, which accepts from the religious heritage only that which can be rationally demonstrated, independently of divine revelation. Nevertheless, although he was not a rationalist, Anselm devoted tremendous effort to the deployment of reason and logic as important instruments in matters of religion. For although reason does not produce faith, although faith precedes reason, reason is essential in order to illuminate the content of the faith one has already accepted: to bring it to its fullness, and thus open the way to God. Anselm was considered, with some justice, as the initiator of scholasticism; he was its driving force. He also devoted much attention to explaining those elements of the Christian faith which, like the doctrine of the Holy Trinity and the Word made flesh, were considered the most unfathomable and mysterious; he tried to lift the veil of mystery surrounding them as much as possible and bring them out into the clear light of day. In his treatise *Why did God Become Man?* he tried to demonstrate that the whole story of the incarnation, the sacrifice, redemption and reconciliation could not have happened in any other way, could not have been conceived by God in any other way except precisely the way in which it took place; that God took the

only logically possible path; and that we should pay no heed to the mockery of infidels who claim that Christians denigrate their God by the story of how He came down to be born of woman, was nourished by a woman's milk, suffered hunger and exhaustion, and finally was nailed to the Cross among criminals – an absurd claim, unworthy of our attention.

Anselm's Ontological Argument has been the subject of much criticism, from a variety of philosophical viewpoints. Here is a summary of some of these critiques:

St Thomas Aquinas rejects Anselm's argument as invalid because he says it *presupposes* that the being than which nothing greater can be conceived actually exists, and that Anselm does not prove this hidden premise. Furthermore, Aquinas objects, Anselm also assumes that one can apply the same sort of reasoning to God that one applies to some truth which is obvious because the predicate is contained in the subject – for instance, the statement 'man is an animal'; in order to do this one must already know the essence of the subject expressed in the predicate, but this is not the case with our knowledge of God: we do not know His essence. So we cannot deduce His existence from it; we can deduce it only from its effects.

Critics of the Ontological Argument often direct their attention to the version of it that is given by

Descartes. According to this version, since God is the most perfect being, i.e., He possesses every kind of perfection, and since existence is a kind of perfection, God's existence is part of His essence in just the same way that it is part of the essence of a triangle that the sum of its angles is equal to two right angles. Therefore God must exist. It is worth considering whether these two versions of the Ontological Argument – Anselm's and Descartes' – really are identical.

Some seventeenth-century critics of Descartes pointed out that there is nothing obviously true in the claim that existence is a kind of perfection, just one kind among others. Kant went even further: according to him, statements which are necessarily true cannot be about existence. The statement that a triangle has three sides is necessarily true in the sense that if a triangle exists, it will have three sides; however, it tells us nothing about whether or not the triangle exists. But defenders of the Ontological Argument include existence in the idea of the most perfect being; consequently, Kant argues, their statement that God exists becomes a tautology. No statement about the existence of anything whatever can, in Kant's view, be an analytic statement – that is, one that is true solely by virtue of the meaning of its words, like 'every triangle has three angles'. Existence cannot be an inherent element of any idea;

or, as Kant puts it, existence is not a true predicate. The word 'is' is a copula between the subject and the predicate; it joins them together. It is not another predicate. When I think about something, I do not add anything to my idea of it when I say that it exists. The question of God's existence remains open when I define God as the most perfect being, or as a being that lacks nothing. A conceptual familiarity with something may, in some cases, be enough for us to be able to claim that that thing is *possible*, but never that it exists in reality.

Some philosophers, perhaps influenced by Kant's objections, proposed a different, reduced version of the Ontological Argument. This boiled down to the statement that if God is possible, then God is necessary. But this is clearly not what Anselm had in mind. Besides, according to Kant, even the possibility of God's existence cannot be proved *a priori*. Schopenhauer pointed out that Aristotle, as if anticipating that something like the Ontological Argument would come along one day in the future, distinguished between questions of definition and questions of existence, saying that they must be set apart.

Bertrand Russell's objections to the Ontological Argument, although differently formulated, are essentially similar to Kant's: to establish the existence of something is to demonstrate that actual

instances of the idea under scrutiny actually exist in the world. No amount of scrutiny of the idea itself can establish with certainty the real existence of any instances of it.

Anselm's theology suggests a number of questions that are worth pondering. Some of them are related to the Ontological Argument.

It seems to follow from the Ontological Argument that if we were to deny the existence of God while knowing the meaning of the words we were using, and in particular the meaning of the word 'God', we would be contradicting ourselves. Aquinas denies that this is so; from his critique it follows that it is possible for an atheist to be consistent. Who is right, Anselm or Aquinas?

Can we maintain that God is a necessary being while rejecting the validity of Anselm's argument? And what would it mean for His existence to be 'necessary' in that case?

Does the feeling of necessity we have when we grasp the meaning of certain sentences result merely from linguistic convention (as in the case of the frequently cited example, 'all bachelors are unmarried')?

What would it mean to say that some object of our thought is better or more perfect if it exists than if it does not?

Anselm also suggests some questions about the nature of God that are unrelated to the Ontological Argument. We may legitimately ask them even without believing in God, for they are questions about logical coherence.

There are some things, Anselm says, that God cannot do. He cannot undo the past. He cannot make the true false or the false true. He cannot lie. Does this mean that He is not omnipotent after all?

God is merciful; He has mercy on sinners. But He is also immutable and not subject to emotions. How, then, is His mercy to be understood? Similarly, if God is just, how can He save some sinners while condemning others, the former by His mercy, the latter according to justice, if the evil done by both is similar?

Finally, there is the most general question suggested by Anselm. He does not formulate it as such, but it emerges ineluctably from his works. It concerns his famous sentence, 'I do not desire to understand in order to believe; I believe in order to understand.' Although Anselm devoted great effort to rationalizing Christian beliefs, he did not try to make faith dependent on the validity of his arguments; on the contrary, he insisted very strongly that faith precedes them all. The question that arises in this context is this: is it irrational to believe in

God if we know that there is no reliable evidence – evidence of a kind that could withstand scientific scrutiny – of His presence? And if it is irrational, what does 'irrational' mean here?

Knowledge, Faith and the Soul:
Is the world good?

ST THOMAS AQUINAS
c. AD 1225–74

No great philosopher has had, in modern times, as many devoted followers and defenders as St Thomas Aquinas. We may call someone a Kantian or a Platonist because he has adopted some particular idea that Kant or Plato famously expressed – for example, the belief that there are *a priori* synthetic statements or that mathematical objects have a real existence, independent of our thought; but we would surely be hard put to find someone who had embraced Plato's or Kant's philosophy *in toto* and adopted it as his own. This is not the case with Aquinas; there have been many people, and among them some truly brilliant minds, who espoused the entire corpus of Aquinas' philosophical work. The reason for this is in large measure institutional, for in the Roman Catholic Church Thomism has

the reputation of a doctrine which better than any other serves to confirm and strengthen the intellectual heritage of Catholic culture. The Dominicans, shortly after the death of their founder, declared Aquinas' teaching their official doctrine (though the Franciscans forbade the reading of the *Summa Theologiae* without corrections), and Leo XIII, shortly after ascending to the papacy, in an encyclical devoted to extolling the virtues of philosophy in the effort to support faith, particularly advocated the study of St Thomas Aquinas, prince of philosophers, as the main instrument of this effort.

Of course, Christian philosophy presupposes the dogmas of the faith, but unlike theology it does not take them as the starting point of its enquiry; it is meant to be an intellectually independent discipline. A distinction is usually made between the order of faith and the order of knowledge, but it was made in a variety of different ways in the history of Christian thought in the Middle Ages. Some thinkers distinguished between the two domains in a way that left them nothing in common; sometimes they even declared that if human reason, when it operates properly according to its own rules, reaches conclusions that contradict the dogmas of the faith, one should not worry about this or try to reconcile the two, but continue to hold both, asserting two mutually contradictory theories at the

same time. Others, in accordance with a tradition rooted in certain currents in Christianity, condemned all philosophy and secular knowledge as sinful amusements and insisted that engaging in them would bring no profit, but only harm the faith. Others still insisted that only those elements of the dogmas of the faith should be accepted which could hold up before the court of secular reason; they, however, clearly do not belong to the Christian spiritual realm, but rather to the history of the Enlightenment.

Aquinas took none of these three paths. He was concerned with the problem that preoccupied all Christian thinkers: since man participates in both orders, the temporal and the eternal; since he has a body, but his chief concern is supposed to be his soul; since he lives in a world of sense-experience, but his proper home is heaven; since he makes use of his faculty of natural reason, but his source of illumination in the most important matters is faith; since he belongs to various temporal collectivities and communities, and is a participant in secular history, but also belongs to the Church, the mystical body of Christ, and is also a participant in sacred history – how are these two orders of man's existence related, and how are they reconciled?

The whole corpus of Aquinas' work has one over-arching aim: to rehabilitate the natural, temporal

Leszek Kołakowski

order (although it will always be subordinated to the eternal order, to God's plans). According to Aquinas, man cannot, contrary to what the Augustinian tradition claimed, be defined as a soul that acts independently of sense-experience; man is a complex being, consisting of body and soul, both of them inherent parts of what we are. The rational soul is the form of the human body (a doctrine officially adopted by the Catholic Church), and this complex unity of body and soul is what full human existence means. The fact that we are corporeal beings is not a minor or contingent matter, the result of chance or a reason for shame; it is part of the definition of our existence. So it is natural that our material existence, and our earthly goods, should matter to us, although spiritual goods must of course always take precedence; material goods are the soul's instruments. The soul gives us everything we have: our body, our life, our intellectual faculties.

Expressions of this position can be found in all the elements that make up the monumental edifice that was Aquinas' intellectual world. Whatever is natural, is good; the natural is an instrumental good existing alongside the highest good, and is not to be condemned or disdained. The two are not separate, nor are they independent of each other; they coexist and together define our life and the life

of the world, although the natural will always be subordinate. Our temporal aims should not be disdained, but arranged in a hierarchy and subordinated to our eternal aims. The created world is good, for God created it; but He did not make it equal to Himself. God makes use of the created world as an instrument for increasing good. The question of why God created the world in the first place, since He could not suffer any lack or feel any need for anything else but was self-sufficient in His fullness and perfection, is also given a beautiful answer by Aquinas: good has an innate tendency to expand, to create more good.

A similar principle – a duality of subordination – is at work in Aquinas' approach to the relation between revelation and the knowledge obtained from the natural world. Natural knowledge is laudable and worthy of acquiring, but it is subordinate to revelation. Faith and reason are not entirely separate with regard to their object; there is some overlap, some common ground, for certain truths which we know through faith can also be justified by the instruments of natural reason, independently of faith. Thus we can put forward rational arguments for the existence of God and certain of His attributes. However, there are many truths of faith that it is beyond the power of secular reason to fathom: for instance, those that concern the Holy

Trinity. The claim that the world had its beginning in time must also be accepted through revelation; its truth cannot be demonstrated with the aid of our natural faculties of reason. On the other hand, it is also clear that revelation does not contain all the knowledge that is accessible to us. Hence we need secular knowledge, knowledge that is independent of revelation, and should not disdain secular science. The subordination of natural knowledge is two-fold. First, there is negative control: when reason, operating according to its own rules, reaches conclusions that contradict revelation, this means that an error has occurred somewhere – reason has not operated properly in accordance with those rules. When reason functions properly, independently of revelation, it will never be in contradiction with revelation. Second, there is the hierarchy of aims: secular reason must always remember that natural knowledge is not an end in itself; it is instrumental only, needed in order to serve God and our eternal destiny.

The relation between secular society and the Church, the guardian of eternal goods, is similar. Here the proper order of subordination is as follows. The state, the secular authority, enjoys autonomy of a relative kind, which the Church does not aim to supplant; it does not demand theocracy. But the Church must oversee its operation in all matters to

do with morality, religious worship and salvation. Thus Thomism does not call for the destruction or condemnation of secular life, but wants Christianity to assimilate it and recognize it as good; to give it its blessing, to see God's plan in it, to see in it a great and relatively autonomous part of the divinely established order. Secular and sacred history are not entirely independent, for although sacred history – the history of creation, redemption and man's salvation – goes beyond secular history in its main points, nevertheless secular history can support sacred history; it can be a bulwark for it. It can, so to speak, carry sacred history's treasures on its shoulders, and expect from it illumination and signs about the ultimate end. In modern Thomism a similar distinction is made, and a similar hierarchy adopted, in the relation that obtains between the idea of the human individual, who is a part of society, and the idea of the human person, who is a participant in the plan of salvation, and in his uniqueness and infinite value goes beyond the bonds and relations of collective life.

Aquinas' rehabilitation of the natural order is also evident in the fact that natural theology – the application of the instruments of secular reason to the truths of the faith – begins from nature, from our sense-experience of the world. Each of Aquinas' five ways to God takes as its starting point truths

that we can establish only by observing nature; they reveal God not in His self-sufficient, self-referring perfection, but in His relation to creation: as a Creator, as the source of motion, as the first cause of everything, as the most perfect of all beings, as the architect of the end-directed order of the universe and, finally, as the only necessary being – for all of creation is contingent and might not have existed. These five ways, or proofs, have of course been the subject of much critique. But the point of this brief glance at Aquinas' mighty edifice is not to analyse their validity but to bring out Aquinas' insistence that this path to the nature of God is the royal path of reason: what we know about God is what we know about the effects of His existence on the world. Each of his ways to God aims to demonstrate how everything in the world participates in the divine order and points us back to natural causality. In elaborating his five proofs Aquinas uses arguments taken from Aristotle, and sometimes from Avicenna and Maimonides. For instance, since there is nothing in God that is not God Himself, nothing that does not exist necessarily, none of our words can refer both to God and to the things of creation, not even the word 'is'; all our natural knowledge comes from perception; since we know various objects in the world, we can also come to know the cognitive faculties

through which we come to know them, and through this knowledge we can come to know ourselves; I cannot have a direct intuition of myself, I can know myself only through my knowledge of the external world; abstract knowledge also comes from perception, because it is through our perception of material things that we can come to discover their non-material aspect – their form.

Aquinas' philosophy is, among other things, a response to those intellectual currents which denied that there was any point in trying to reconcile secular knowledge and faith, and in general to the various claims of secular life, human reason, morality in temporal concerns, civil society and civil authority – claims to independence from faith, from the Church and from God. One important theological ingredient of these was the claim that divine Providence, and even divine omniscience, does not extend to events in the future, which depend on chance or on free human will. For Aquinas, however, divine Providence, omniscience and wisdom are all-encompassing; God's guidance overlooks nothing. It follows that no domain of secular life can be entirely independent; the independence of secular domains is merely relative, and in matters of truth and moral qualities they are subject to divine norms.

A question that naturally arises, and has always

arisen, from this faith, is the question of evil: the problem of the presence of evil in the world. Aquinas on the whole accepts Augustine's answer: existence as such is good, and evil is not being but absence, or lack, which God allows in order to bring about greater good through it. Evil is the absence of what should be; it is not an evil for man that he does not have wings, but it is an evil for him to lack a hand. However, Aquinas does not share Augustine's notorious belief in the ubiquity of moral evil, which seeps through into every aspect and domain of our existence. Nor does he share Augustine's view that after the Fall the human will is capable only of doing evil unless guided by God's gratuitously bestowed and irresistible grace. Aquinas believed that each of us is where he should be in the order of being; we all have our allotted place. He does not seem much interested in the demonic side of human existence.

There can be no doubt that Aquinas was one of the most powerful pillars of European philosophical culture. But philosophy has changed so much since the thirteenth century, in its language and the kinds of questions it poses, that although there are many Thomists in the world, Thomism is no longer a significant inspirational force or stimulus in philosophy outside Thomist circles.

Here are just three out of the thousands of questions that Thomas Aquinas' work suggests:

Can the claim that existence as such is good have any other or additional meaning than the assurance that everything was created by God?

If good has a natural tendency to expand (which also explains why God created the world), does the same apply to evil? And what would be the consequences of such a doctrine?

The objects of our sense-experience are contingent: they might not have existed. Thus there was a time when they did not exist. They must therefore have been created by a being that must always have existed and cannot not exist – a necessary being, therefore, and thus a divine one. Is this reasoning sound?

What There Is:
Do ideas exist?

WILLIAM OF OCKHAM
AD 1285–1347

William of Ockham, or Occam – or simply Ockham,
for that is how we generally refer to him, forgetting
about the English village where he is supposed to
have been born – is famous chiefly for two things,
closely bound up with each other. His first title
to popular fame is his radical and multi-faceted
defence of the nominalist doctrine. The second is
the instrument known as Ockham's Razor. This
latter is a principle most often heard in a version that
is not entirely accurate, although the formulation is
in accordance with the spirit of Ockham's philosophy: 'Entities should not be multiplied unnecess-
arily.' The exact formulation is: 'Multiplicity should
not be posited beyond necessity' ('*Pluralitas non est
ponenda sine necessitate*'). The Razor is a general
principle of reasoning similar to the principle later

called the law of economy or the law of parsimony, present in a variety of guises in modern philosophy. It is a principle that attacks certain branches of scholastic philosophy and theology which peopled their world with unnecessary constructs and sometimes produced various not very fruitful theological, and also physical, laws. Ockham was not the first creator of this doctrine; it was invoked by Aquinas as an example of an argument – the falsity of which he went on to demonstrate – for the non-existence of God: that God is unnecessary if the world can be explained without Him.

Ockham was interested in many things, and was probably the most brilliant and most important thinker of the fourteenth century. He was, among other things, a masterly and extremely subtle logician. Here, however, we are interested only in questions that concern nominalism and its consequences. His anti-papal writings we will leave aside, as we will leave aside his part in the Franciscan debate about poverty (for Ockham was a Franciscan friar).

Nominalism is a philosophical position that can be presented either as a metaphysical doctrine or as a principle of reasoning. Its exponents claim that all general terms designate only the individual objects which we call by such names, and do not refer to anything else. Thus the object designated by the

name 'man' is any individual man, and the object
of the name 'leaf' is an 'individual leaf', and 'yellow'
describes any object that is yellow. We do, of
course, use abstract terms, but the things that
correspond to them in the world are not general
or universal entities of any kind (either existing
independently or inherent in things), but concrete,
specific things. There are no such things as 'universals', the abstract entities which people the discourse
of philosophers.

The controversy about universals has continued
almost without interruption throughout the history
of philosophy, from the ancient Greeks up to the
twentieth century (Bertrand Russell, for instance,
thought that nominalism was an untenable doctrine, while the Polish philosopher Tadeusz Kotar-
binski was a radical nominalist). The various senses,
both philosophical and ideological, of the term
'nominalism' and its opposite (called 'realism' in
one of the many meanings of this term, and sometimes identified more precisely as 'conceptual
realism') were numerous, as were the contexts in
which nominalist or anti-nominalist positions were
expressed, but here we will concentrate on the
classic fourteenth-century version. According to
nominalism, generality is a feature of language, not
of things. If things are similar in some of their
aspects, for instance all yellow things, this does not

allow us to infer the existence of yellowness as a separate entity or to say that yellowness inheres in yellow things as their common nature or essence. Such expressions lead only to vain speculation which in no way increases our knowledge. What we really know is based on direct experience, and the things we experience directly are twofold: the things we can see and touch, and our own mental activity, i.e., seeing something or wanting something. But there are no general entities in experience, although of course we tend most often to express our knowledge of the world in general terms, when we say, for instance, that man is an animal or that a poplar grows more quickly than an oak tree. Our intellect can grasp the general in the sense of being able to observe and retain the common features of things, the resemblances between them; but our grasp of those resemblances tells us nothing about the existence or non-existence of anything. Belief in universals is a misleading multiplication of the only reality about which we know anything – the reality that is accessible to our sense-experience. The traditional distinction between existence and essence also makes no sense: for it is not the case, clearly, that there is something that doesn't exist and to which existence is merely tacked on. Similarly with the belief that the general precedes the particular: for if the general existed

before the particular, it would itself be a particular entity, which would make it a contradiction in terms.

The belief that direct experience is our only reliable source of knowledge about the world has theological consequences. We do not know God through direct intuition, the way we know the physical world; the physical world is simply a fact, and like all facts it is contingent. Neither Anselm's Ontological Argument for the existence of God nor Aquinas' proofs of His existence are valid. Aquinas' proofs are all supposedly based on experience, on what happens in the world. But there are no degrees of perfection or final causes in our direct experience. So it would seem that, according to Ockham, natural theology – that area of philosophy which deals with theological matters – is entirely discredited.

Ockham certainly upheld the Christian faith as a set of divinely revealed truths, but he did not believe that these truths could be proved by reason. Consequently, his philosophy opens up a deep rift, an unbridgeable chasm, between natural knowledge and faith. We may legitimately believe that there is such a thing as regularity in nature – in other words, that although each fact is contingent, sequences of facts arrange themselves into certain regularities; but these regularities are not laws and cannot be

absolutely certain, because God could interrupt or revoke them at any moment if He chose. God is absolutely omnipotent: He can do anything that is not self-contradictory. The world as it is has no necessity in and of itself; it is how it is because God's free will made it so, and that will could have been entirely different.

Man, too, has free will: it is only because he has free will that his actions can be good or evil. But if so – if human actions are not causally determined, but depend on our free will – how can God know about them in advance? Yet how can He not know about them, if He is omniscient? Ockham devotes a whole separate treatise to this problem, and the complicated argument it contains seems to conclude that although both kinds of truth – about God's omniscience and about human free will – must be acknowledged, we will never, in the course of our lives on this earth, be able to understand how they can be reconciled. The truths of revelation must simply be believed; revealed truth is not obvious to human understanding. Accepting absolute divine omnipotence, restricted only by the law of non-contradiction, entails consequences that can seem disturbing, perhaps even terrifying. God gave us moral commandments, but they are His arbitrary decrees; had He wished, He could have decreed the contrary of the moral laws we know from the Ten

Commandments. Then it would have been our duty to perform the contrary acts, and fulfilling the Ten Commandments would have been a sin. Thus we may conclude from Ockham's argument, although he does not explicitly say this, that God does not command or forbid something because it is good or evil, but the other way around: our deeds are good or evil solely by virtue of God's decree, good if He commands them and evil if He forbids them.

The problem is, of course, a traditional one, and Ockham was not the first to argue from the premise of God's omnipotence to the conclusion that moral laws are established by divine decree. Ockham's God is, in fact, none other than the God of St Augustine. Another consequence of God's omnipotence is that He has no moral obligations towards His human subjects; we cannot expect that someone who has scrupulously fulfilled all God's commandments will inevitably be saved, or that the incorrigible sinner will inevitably be damned. Everything depends on God's arbitrary and irrevocable will, which owes obedience to nothing and no one and abides by no rules.

It is worth mentioning that the same premises about God and natural knowledge – the presuppositions concerning experience as the only source of natural knowledge and God as absolutely free and omnipotent – led certain nominalists to draw even

more radical conclusions. Ockham does not perhaps explicitly embrace them, but he does sometimes seem to suggest them. Nicholas of Autrecourt, a Parisian nominalist of Ockham's day, was condemned as a heretic for a variety of philosophical opinions, some of them similar: fifty-eight 'errors' were singled out as heretical. Among them was the claim that we can have no certainty about any substance except our own soul; that apart from the truths of faith, the only truth of which we can be certain is the law of non-contradiction; that we do not know if there are any effects that have natural causes, for God can be the cause of everything; and that there is no way to demonstrate the reality of what we perceive. Ockham, too, was suspected of heresy, and was summoned to the papal court at Avignon, where his case was lengthily examined. In the end he was not condemned as a heretic (though he was excommunicated – not, however, because of his philosophical errors, but because of his attacks on the Church) and fled to Munich, where he availed himself of the protection offered by Ludwig of Bavaria. Radical nominalism and empiricism, of which Ockham was then the most brilliant exponent, ultimately led, if taken to their logical conclusion, to the belief that the only things we know directly are our own perceptions; we cannot know whether anything else really exists,

for God may have caused us to imagine it; He may have arranged things in such a way that none of the things we think exist really exist. Nor do we have any proof of natural causality; and there is no reliable natural theology, although there is revelation, which we must trust.

In order to understand fourteenth-century nominalism, we must remember that the nominalist doctrine is considered a significant source of the Reformation in the sixteenth century. Luther was a student of nominalists. The connection is this: nominalism created a rift between religion and reason, between divine matters on the one hand and secular life and knowledge on the other; it maintained that all things and events in the world are radically contingent and that they are nothing in themselves, for they are the work of the arbitrary will of the Creator. Similarly with moral commandments: they are simply decreed by God, and we must obey them because He commands us to, not because they are good in themselves. Holy Scripture is the only *summa* of knowledge about divine matters; theological and philosophical speculation is utterly worthless and contemptible, and adds nothing to our knowledge in this sphere. And since we have revelation, we have no need of any other support; in particular, we have no need of Church tradition as a separate source of knowledge. We

must bow humbly before God and His Word, without demanding recognition for any alleged merit of our own in the hope of salvation.

But the connection between nominalism and the Reformation, though it does exist, is not a straightforward one. The writings of John Wycliffe, for example, an Oxford philosopher of the second half of the fourteenth century, were another important source of Reformation ideas, and Wycliffe was an opponent of nominalism, for theological reasons (because nominalism contradicts the dogma of the Holy Trinity) and philosophical ones (because in true statements we say something about things, so predicative expressions, or general names, must refer to a corresponding reality), as well as moral ones (because nominalism denies the existence of a common human nature, and thus encourages egoism). The history of how various doctrines in theology and philosophy influenced one another is extremely tangled and complex.

Here are some of the questions that Ockham suggests:

According to the principles of nominalism, what (its critics asked) is, for instance, a Chopin piano concerto? Is it a piece of paper covered with musical notation? Or is it, perhaps, an event that occurred in Chopin's mind? Or is it every particular instance of its performance?

Does God's absolute omnipotence really entail the consequence that all the moral rules He revealed to us are His arbitrary decree, and that it makes no sense to say that they are good in themselves, independently of being decreed by Him?

Let us assume that God, in His omnipotence, is causing us to imagine everything we experience and think of as real, and that the world of our perception is an illusion. What would be the difference between this world and a real world identical with it in content? How could the reality be described so as to distinguish it from the illusion?

God, the World and our Minds: How can we achieve certainty?

RENÉ DESCARTES
1596–1650

Is there anything we can know for certain? With a certainty that admits not even the shadow of a doubt? Descartes was not the first to tackle the question, but the nature of his reply made him one of the great masters of European philosophy, and caused what we might call a cultural mutation. Even while he was still alive, and certainly after his death, from the second half of the seventeenth century onwards, every philosophizing mind had to define itself by reference to Descartes, whether its attitude to him was approving, critical or fiercely hostile. He became a focal point, a fundamental point of reference, as unavoidable as Kant was to be in the nineteenth century. Descartes is considered the initiator of the process which was to ripen into the Enlightenment, although he diverged from the

classical French Enlightenment in so many respects that at first glance it is hard to see any connection between them.

Why should we desire such certainty? Descartes' first publication might seem to suggest that we need it for practical purposes; it will help us gain mastery over nature and thus increase our medical and technical skills. To this end we must emulate those sciences which have already achieved the greatest certainty, namely mathematics. But it soon becomes clear that the certainty Descartes is concerned with is metaphysical certainty: a certainty that would encompass, without the risk of failure, our knowledge of all reality – God and the universe, matter and the soul. The aim was to find the absolute beginning of knowledge, the starting point that is immune to error and doubt. And since we do sometimes fall into error, we must, he thought, begin by investigating the causes of error.

When we dream, Descartes observes in his famous and awe-inspiring argument, our dreams seem real to us; could it be, then, that what we perceive when we are awake is also a kind of dream, a picture with no corresponding reality? Perhaps a malicious demon is deceiving us, misleading our senses with false impressions, leading us astray for reasons best known to himself? We think our impressions are real, but they may be just as illusory

as our dreams. Is there any way to escape the clutches of his hypothetical deceiver? Indeed, there is: for there is one thing the deceiver cannot deceive me about, and that is the fact that I am experiencing something – perceiving something, thinking something. I exist, therefore: for here I am, thinking and perceiving. On this point I cannot be wrong; here no demon, however powerful, could be leading me astray.

This insight, expressed in the form 'I think, therefore I am', became one of the fundamental ideas which characterized modern philosophy. The 'therefore' was later eliminated, because Descartes did not want the phrase to be construed as an inference, a syllogism without a major premise; it was not supposed to be a deduction, but rather one indivisible, compelling act of intuition: 'I think, I am'. The word 'think' is used here in a very broad sense: it encompasses all mental acts and all experience, both the activity of the will and the activity of thinking in the strict sense.

What can we do with this discovery and what follows from it? A great deal, according to Descartes. When I consider my mental acts, I immediately notice that they, and they alone, constitute my essence; no other kind of thing, and in particular not my body, is part of my essence. From this I can infer that the soul, a thinking thing, does not depend

on the body for its existence and can therefore exist without it. In this way we can demonstrate the immateriality and, after some further steps in the argument, the immortality of the soul, in accordance with the Christian faith.

But this is not all. When I consider my own nature, my conscious, thinking self, my soul, in other words, I find that it contains the idea of the most perfect being – the idea of God. Since I myself am not perfect (the very fact of my doubting is proof of this, if proof were needed), I could not have come up with such an idea on my own, so it cannot be a figment of my imagination. But it must come from somewhere; I can only conclude, therefore, that it must have been implanted in my mind by the very being of which it is the idea, namely God. Therefore God exists. (Another of Descartes' arguments for the existence of God was akin to Anselm's ontological proof: existence is a kind of perfection, so the most perfect being must exist, by definition.)

God's existence leads to conclusions which can form the basis of our trust in knowledge, in the possibility of natural cognition. The malicious demon cannot deceive me about one thing at least: my conscious experiences, the fact of my existence. But God, being absolutely perfect, cannot deceive me at all, because deceit cannot coexist with perfec-

tion. We can therefore, once we have assured ourselves of God's existence, safely trust our instinct that our world is real, not a dream or an illusion, and that everything we perceive clearly and distinctly, all the things that seem obvious to us, are really true. Now we can trust our knowledge and overcome doubt; we know what we may accept as true and what kinds of things we may doubt.

But (critics objected) we are often mistaken, and God allows this; He does not protect us from error, does not nourish us with absolute certainty. Descartes acknowledges that of course we are often wrong, but this, he says, is our own fault, not God's; there are innumerable things about which there is no certainty, and as long as this is the case we ought to suspend judgement about them. If we none the less believe something that reason forbids us to believe, we become victims of error through our own fault – by an act of will that goes against reason. We are able to do so because God endowed us with free will – a precious, extraordinary, miraculous gift, but one which we can also turn to bad use, which is what we do when we believe something on insufficient evidence, assent to something that is not clear and distinct to our reason.

Thus it was that Descartes arrived, or thought he had arrived, at the absolute and compelling starting point for human knowledge, one that was quite

independent, in his view, of past philosophical tradition. For this tradition, particularly for scholasticism and for the heritage of Aristotelianism, he had little use. Unlike Aquinas, Descartes did not believe that we could acquire knowledge of God from our knowledge of nature; there was no way from sense-experience to God. The only way was through the idea of Him that is in our mind.

Of course, not everyone was convinced by his arguments. Both Catholic thinkers and rationalists of various hues pointed out the mistakes in his reasoning. You cannot, they said, appeal to your sense of obviousness in order to prove the existence of God and then proceed to make God the guarantor of that criterion of obviousness, for then you are caught in a vicious circle. Furthermore, from the fact that I cannot see how anything other than thinking could be a necessary constituent of my conscious acts it does not follow that those acts cannot be dependent on my body. Moreover, God's goodness is no guarantee that He cannot deceive us: He might want to deceive us for our own good, just as adults sometimes hide things from children for their own good. And do we really know for certain what 'I exist' or 'I am' means?

Descartes rejected these objections. He was unshaken in his conviction that he had found an infallible signpost, indeed, an infallible foundation

on which the whole edifice of knowledge could be rebuilt, relying on reason alone. In history he had little interest, whether it was the history of ideas, full of absurd superstitions, or natural history. About natural history he said that we can always reconstruct it mentally for ourselves, in order to gain a better understanding of the world, but we need not worry about the truth of that reconstruction. The world according to Descartes is, of course, governed by inflexible laws established by God's will. They were decreed by God and could have been different; there is no necessity inherent in the laws of nature. Even the truths of mathematics were created by God's decree; He could have made an entirely different mathematics if He had wished, however much of the mathematics He did choose may appear necessary to our reason.

It is not surprising that the Cartesian edifice, in which critics found so many holes, inspired a large variety of doctrines, some of them mutually contradictory, all of them important in the history of our intellectual culture. There were some attempts to yoke Descartes to the Christian tradition, but it seems fairly safe to say that they were not successful. Catholic philosophers usually tried to demonstrate that he was the source of the fundamental errors into which modern thought had fallen.

Descartes did, it is true, suggest that the only

things we can know directly are the contents of our minds: our perceptions and mental experiences. And even though he went on to prove (as he thought) that these mental experiences reflect a corresponding reality, he was forced, in order to achieve this, to appeal to God's truthfulness, and left behind him the problem of the so-called bridge: the problem of how to get from our perceptions to reality. Thus, willy-nilly, he created the suspicion that the world is a figment of our imagination and thereby opened the way to idealist philosophy, of which Catholic thinkers took a particularly dim view.

Furthermore, since Descartes believed that the world was ruled by the implacable laws of mechanics, he seemed to be denying the possibility of divine intervention. This was in stark contradiction to the traditional belief that the hand of God was at work everywhere and that His intention could be seen in everything: in punishments, in rewards, in warnings, in reprimands and in praise. Descartes did not explicitly deny miracles, but if one accepted the Cartesian world, one could not really believe in them. His world was a superb machine, not an endless succession of independent revelations. The body, too, was a machine, operating according to the same laws of mechanics. Although it might have seemed that Descartes attributed more than others to God's rule – that, like some nominalists, he made

everything, including the truths of mathematics, dependent on God's decrees – it turns out that we can study the world without any reference to God; and just as we may forget about God in studying the world, so we may forget about the immortality of the soul when studying the body.

Thus Descartes was accused not only of idealism, but also of godless materialism and of arrogant over-confidence in reason, which would pronounce about everything by itself. He rejected, it was said, both divine Providence and the dogmas of the faith; and his occasional assurances that he was a loyal Christian, obedient to the Church, were disbelieved and attributed to prudence rather than sincerity.

Here are a few of the puzzling questions that almost every sentence of Descartes' writings suggest:

Let us assume that a malicious demon really is deceiving us and that everything we believe to be real is mere illusion; perhaps the world does not exist at all, or perhaps it is quite different from what we think. Would this make any difference (assuming that our perceptions are unchanged)? Furthermore, if we assume (this is not Descartes' question, but it could be helpful in trying to understand his thought) that the world was created just a minute ago, together with our memories and everything we take to be evidence of the past, would

that make any difference to our lives or to the way we think? And why should we find such an assumption disturbing?

Would it be true to say that my existence is the same thing as my consciousness of my existence?

And finally, if the truths of mathematics really are arbitrarily decreed by God, what does it mean to say that they are true?

The Nature of God:
Do we have free will?

BENEDICT SPINOZA
1632–77

Some have called Spinoza godless or an enemy of God; others – the majority – saw him as a pantheist; others still thought of him as a deeply religious man, a man possessed by God. One could muster arguments to support each of these views. One of the reasons for such wide disparities in interpretations of Spinoza is, paradoxically, the fact that he aspired to the greatest possible precision in expounding his thought. In order to achieve this he arranged his main work, the *Ethics*, in 'mathematical order', in the style of Euclid and other mathematicians; it contains axioms and definitions, and every statement in it is supposed to be deducible from those axioms. The reader will soon notice, however, not only that various statements, for instance observations about human psychology, are artificially

tacked on to the axioms, but above all that metaphysical concepts – in contrast to simple ideas
of geometry, which are intuitively understood –
do not become any clearer thanks to this so-called
deductive method of exposition. So in this very
brief glance at one of the aspects of Spinoza's philosophy we will ignore this arrangement, even though
for Spinoza himself it had great philosophical significance, for every detail of it was supposed to be
applicable to the whole, and this, for Spinoza, was
the right way not only to think and do philosophy,
but to live. Indeed, Spinoza's reflections about ethical concerns are the culmination of the *Ethics*: perfection and happiness, the good, man's freedom
and his enslavement to the passions. But the most
influential for later philosophy, and the most contro-
versial, was what Spinoza had to say about God
and His relationship to the world, and about the
human soul. This is also the most involved and
strangely tangled part of the *Ethics*.

God's existence is logically entailed by His
essence; in other words, He exists by definition.
Here Spinoza follows in the footsteps of St Anselm
and Descartes: God cannot be conceived of as not
existing. He is also His own cause: He can have no
cause outside Himself. He is infinite, in the sense
that nothing can limit or affect Him. So far, all this
is in accord with Christian theology. However,

other attributes of Spinoza's God are less so. God and Nature are one and the same. God is also called substance; and there can be only one substance. There is also another sense in which God is infinite, namely that He is endowed with an infinite number of attributes; however, we only know two out of this infinite number, and they are extension and thought. For Descartes, substance is of two kinds: extended or thinking; the soul is a thinking substance, and 'thinking' is a general name encompassing all mental acts. Spinoza considers both thinking and extension to be attributes of God: God is extended, and God is thought. God's thought is not, however, a psychological process, like human thought. What, then, is it?

Spinoza's answer is part of a complex larger picture. Particular things, including humans, are not finite, independently existing God's creatures, nor are they parts of God. They are, in Spinoza's terms, modifications of substance: they express God. Substance itself – that is, God – has no parts. Infinite extension is indivisible: it cannot have parts, for whatever has parts cannot be perfect. And we know that substance has extension, for things are extended, and things cannot be conceived other than in relation to God, so God, too, must have extension. And since human beings think, our thought must also be an expression of God's perfection.

Thus God does think, but His thinking is not a collection of separate thoughts, like ours; it is eternal and all-encompassing Reason, in which we too participate by our thinking, though imperfectly. Our mental acts take place in time, but God is beyond time; if He existed in time, He could not be fully perfect, for He would have to mediate His relation to Himself through memory of what was past and anticipation of what was to come.

What is the picture that emerges from this of the relation that obtains between the divine Absolute and finite things? They are not parts of God, but nor do they have an autonomous existence; and God cannot be conceived by reference to them, only through and in Himself. This is an odd construction, and it is clear from Spinoza's letters that he was conscious of having failed to deal satisfactorily with the crucial question of the relation between God and the existence of things. The universe that we know through experience is not God; we cannot learn about God through experience, but only by conceptual analysis. Sometimes it seems as though, for Spinoza, God is in fact the only being, though he does not say this in so many words.

It is not only we – human beings – who participate in the attributes of substance that are known to us; all things participate in them. Every particular thing is therefore both body and idea. Its being an

idea does not mean that it has any mental life of its own; it is an idea in the sense that it is thought by God, it is part of God's eternal thought. Idea and body are in reality one; whatever happens in one happens also in the other. This is not, however, because body and its idea can causally affect each other in any way; there is also no causal relation between the human body and the soul, although they, too, are one and the same. Even the ancient Hebrews, says Spinoza, vaguely perceived this when they said that God, divine reason and the things it thinks are one and the same.

God's spirit suffuses the universe, and God causes everything that happens in the world of our experience. Not, however, in the sense of bringing about each event by a separate decree, but rather in the sense that in the infinite chain of events each event is implacably determined by preceding events. Causality in the world is absolute: nothing happens by accident. Free will is thus no more than a super-stition of the common folk. Neither God nor man is free in the sense of being able to consider various courses of action and choosing one of them. It sometimes seems to us that we are free in this way, but we delude ourselves: we have no more freedom than a stone. If a stone could think, it might imagine, when falling off a cliff, that it was falling of its own free will; that it had freely chosen to fall. Our

illusion of freedom is no different from the stone's.

But can we, then (someone might ask), punish people for their misdeeds, if everything is entirely determined and no one freely chooses to do what he does, but is governed by implacable necessity? Spinoza says: yes, we can. Just as we kill venomous snakes without asking if they have free will, so, in the name of the common good, we must punish offenders.

God acts from necessity, and that necessity flows from His nature; His actions are absolutely perfect, like Him. And there are no final causes in nature, for their existence would entail the existence of free will.

Many have found Spinoza's metaphysics gloomy: a world in which everything, including everything we do, is absolutely dependent on some irresistible impersonal force which, though it bears the name of God, seems to have little in common with the God of Christian faith, or indeed with the God of any other religion: a God to whom one can pray. But Spinoza thought that his doctrine not only revealed the truth but prepared us for the best sort of life and provided infallible moral guidelines. These are very similar in spirit to the teachings of the ancient Stoics: knowledge about God and the nature of our connection with Him, once we have acquired it, is supposed to protect us from the blows

of fate; once we know that everything is the result of God's eternal and necessary decrees and nothing depends on us, we will be indifferent to the misfortunes that might befall us. We will be free from hatred and anger, from envy and contempt towards others. Spinoza urges us to be content with what we have and help our fellow men. His aim is to steer us towards freedom: freedom not in the sense of some superstitious belief in free will, but in the sense of freedom from evil emotions, from the slavery of the passions. Spinoza really believed that reason by itself can endow us with freedom and happiness. And the good life brings rewards, because virtue, according to the Stoic principle, is its own reward; it is not rewarded by benefits that flow from any external source.

Although God does not reward us by granting our whims and accidental desires, nevertheless, if we are guided by reason, and once we have grasped the truth about ourselves and the world, a love of God will spring up in us. This love, however, is of a singular kind. It is not ordinary love, but the kind of love that nothing can weaken or disturb: an intellectual love. And it is eternal, for our souls, too, partake of eternity – since they partake of God. This eternity is not to be understood as the immortality that religious faith promises us (for there is no memory in it), but as immutable participation in

an immutable God. Intellectual love for God is the same love with which God loves Himself. And this love, once we have achieved it, will be our greatest happiness and our salvation. The common people, however, cannot attain it; to the extent that they obey moral laws, they generally do so only through a miserable fear of hell.

From these remarks we can infer some things about Spinoza's attitude towards the religious communities of his day. When he was expelled from the Jewish community for his unorthodox views, he was a young man with rabbinical training; after his expulsion he joined no Church or sect, but remained without a religion, like his friends, even if they formally belonged to religious communities. He knew the Jewish faith better than any other, of course, but had no respect for it; the Old Testament was an important historical source for him, but he did not believe it was in any way divinely inspired. His attitude to the Church of Rome, evident in his reply to a letter from an old student, a Catholic convert, was one of disgust, although he probably did not know as much about it as he should have. Nevertheless, he was one of the most illustrious advocates of religious toleration; when his landlords, members of a peaceful and tolerant Mennonite community, asked him whether he thought their religion was good, he said yes, it was good and

they should stick to it. For although in his writings he urged us to strive towards a mystical sort of union with God, towards intellectual love and philosophical reconciliation with whatever fate should bring, he knew that his advice was meant for only a very select few, and that the rest, the common folk who are incapable of subordinating their lives to reason, would still need advice about how to lead a good life – the kind of advice that religion provides. This was fine as long as the religion was one that preached peace and unity, and did not breed fanaticism, hatred or despotic government.

In the whole history of philosophy there is no figure as lonely as Spinoza.

Here are some of the questions he suggests:

If we maintain that there is no free will and that everything is ruled by the implacable, undeviating force of natural causality, can we also maintain, without self-contradiction, that moral good and evil exist in the world of human affairs?

Can we really love God if we know that He is an impersonal force, acting from the necessity of His own nature? And what does it mean to say of an impersonal God that He is good?

If we know the causes of our emotions and passions, do those emotions and passions disappear? For instance, will sadness disappear, as Spinoza claimed, if we know its source?

God and the World:
Why is there something rather than nothing?

GOTTFRIED WILHELM LEIBNIZ
1646–1716

Leibniz is often said – I do not remember who was the first to say this of him – to have been the last man in Europe who knew everything. He absorbed the entire intellectual culture of his age, and made significant original contributions to every field of enquiry he studied: mathematics and physics, philosophy and theology, logic and jurisprudence, history and geology. His brilliant mind encompassed everything; whatever he touched was enriched by him. Later such omniscience was no longer possible; the various disciplines evolved too quickly, and very soon there was too much knowledge for any one mind to absorb.

It is not the diversity of Leibniz's interests that is important here, however, but his search for one common root of all the branches of knowledge: his

dream of discovering a common guiding force and ultimately understanding all those branches as parts of a single edifice, constructed according to a single architectural design.

Such a view and understanding of the world was possible, he thought, as long as we obey and constantly keep in mind two supreme principles of reasoning. One of them, generally uncontroversial, was the law of non-contradiction: the principle that anything which contains a contradiction or leads to a contradiction must be false. The other was the principle of sufficient reason – a principle which seems to be a metaphysical rather than a purely logical one. This states that for everything that is real there must be a sufficient reason for its being what it is – a reason sufficient to have caused it to be the way it is and no other way. Every true statement must also have such a reason, or justification. This does not mean that such reasons will always be accessible to us; mostly they won't be. But it is vital to know that such reasons must exist, in every case; otherwise true knowledge is impossible. The kind of sufficient reason we want can be discovered in eternal truths, truths which are necessary and whose necessity we can grasp: the axioms of logic and mathematics. These truths – the truths of reason – were not arbitrarily decreed by God; they could not have been different if God

had decided otherwise. They are true independently of God. God could not have caused it to be the case that 3 multiplied by itself is not 9 but something else. Nor could He have created a precise number corresponding to the square root of 2. The truths of mathematics belong, eternally and immutably, to the world of numbers, and even God cannot change them. This does not mean that He is bound by any external constraints, subject to some other force or law; these truths are identical with God Himself. They are, so to speak, immutable elements of His being; they are not His whims.

But there is also an infinite number of contingent truths: truths about the world and particular events in it. These are truths of fact. Not being necessary, they do not have reasons in the same sense; their truth does not flow from their content alone. That girl may have fair hair, but I can imagine her being otherwise; the supposition that she has dark hair does not lead to a contradiction, as would, for instance, an attempt to deny the proposition that all the radii of a circle are equal. But the principle of sufficient reason states that a sufficient reason must exist for all truths, even contingent ones. However – and this is the crucial point – giving a reason for some event or for some statement from which a given claim follows is not the same as giving a sufficient reason; for a sufficient reason is an ultimate

reason. Explaining a contingent event by some other contingent event won't do; it will not lead us to a reason that is not contingent, however long we make the chain of explanations. What we are looking for is not an efficient cause, which still needs a further cause to explain it, but a cause which itself has no cause – an ultimate, final cause. Leibniz famously asked: why is there something rather than nothing, since nothing is easier than something? If we accept that this question is meaningful (and some philosophers deny that it is), there is only one answer: God. God is the cause of everything, for there are no isolated events in the world: everything is bound together into one mechanism and requires only one final cause. The world is ordered towards an end, and although we are unable to encompass the infinite succession of events and perceive that end – that requires an absolute mind – nevertheless we know that such an end, such a final cause, exists. And since for Leibniz the principle of sufficient reason is a logical principle, it would appear that the existence of God is a truth of logic.

Crucial consequences follow from the claim that God is the only final cause of all events. The world is an infinite collection of monads. A monad is a thing that is absolutely simple, without extension and without parts; it cannot be born or die naturally. Since there are complex things, simple things must

also exist. And a simple thing cannot be affected by other things. Monads are utterly impregnable, immune to all external influences and stimuli; they have no windows, as Leibniz said. This does not mean that nothing in a monad ever changes; changes constantly take place in them, but for internal, not external reasons. Each human soul is a monad, and each human body is a collection of monads adapted to the form of the soul. No two monads are identical; each monad is different from all other monads. There is a hierarchy of monads, at the top of which is the highest monad – God. Thus the universe consists of mental units. But spatial relations are not merely an illusion. They do not have an independent existence; they do not exist in the same sense as the human soul. They are projections. But they are indispensable, for it is thanks to them that there is a common world of human experience; without them we would have no way of knowing that our perceptions refer to the same reality as other people's perceptions.

But how is it possible that monads and things should be unable to affect one another, and yet that all things in the world should be tightly bound up in one immense mechanism? It is possible thanks to what Leibniz calls pre-established harmony. This is a harmony, established by God, of the movements, actions and perceptions of all monads; thanks to it

everything happens as precisely as if by the operation of causal laws and relations, although no such laws or relations obtain. Each monad goes on endlessly, infallibly following the implacable destiny which is part of its nature, and the destinies of all individual monads are arranged into a perfect, harmonious whole by the will of God. So not only does all of this happen as if natural causality were at work in the world, just as we imagine that it is, but the harmony which governs the activity of the monads is so perfect and designed with such exquisite precision that each individual monad reflects the fate of all the others; each is an infallible mirror of the whole universe, its past, present and future, into infinity. This does not mean that we can in reality come to know every detail of everything in the world just by examining our own souls. This knowledge is present in us, but our perception is not sufficiently perfect to allow us consciously to perceive all events, past, present and future, that occur in the world. Our perceptions come in varying degrees of clarity and varying degrees of familiarity and connection with their objects. Only God, timeless and perfect, has total knowledge of everything. He knows that in every true statement some attribute or quality is predicated of an object, and that the relation between them is necessary, so that every true statement is in fact a statement of identity

— an analytic statement, as we have called it ever since Kant invented the description. Every subject already contains its predicate — everything that can truly be predicated of it; the concept of each thing contains its entire, infinite destiny. The true statement which predicates of me that I have just overturned my wine glass and spilt the wine is no less necessary an element of the concept of me than the truth that every triangle has three angles is a necessary element of the concept of a triangle. This latter is a truth of reason — one of the truths we know to be necessary; the first is a truth of fact, a contingent truth. But it, too, turns out to be necessary, just as necessary as any truth of reason, although one would have to have God's perfect intellect in order to grasp this fully.

At this point a familiar question naturally arises; if the fate of the world and each of its elements is already determined in every detail in God's pre-established harmony, can we still claim, without contradiction, that we have free will?

Yes, we can, Leibniz replied; for human notions are not necessary despite being determined in God's scheme. Only those truths are necessary whose negation leads to a contradiction — truths of reason, like the proposition that every triangle has three angles. Contingent truths, truths of fact, are foreseen by God, but for us their negation is still poss-

ible without contradiction; there would be nothing contradictory in saying that I did not, in fact, over-turn my wine glass. We always act in the way that seems most appropriate to us, and we do so freely.

Of course there is evil, sin and suffering in the world. Evil arises from the very imperfection of God's creatures; God could not have created them perfect, for then they would have been His equals, and it is logically impossible for there to be more than one God. God, Leibniz says – in accordance with traditional Christian theology – does not do evil, He merely allows it. But – and this is a crucial point – God allows only the minimum of evil that is possible in the world. God must be infinitely good because He is perfect, and goodness is entailed by perfection, by definition. He is also infinitely wise, and from those two attributes, goodness and wisdom, it follows that He must have created the best of all possible worlds – that is, a world in which the sum of good is as great as it could possibly be and the sum of evil as small as it could possibly be. God looked over all possible worlds and must have solved an infinitely complex equation which produced our world as the solution. A world with no evil in it may have been possible, but it would have to have been a world peopled by automata without free will; God calculated that our world, with all its evil and suffering, still produces infinitely

more good than that other possible world without evil and without freedom. We live, therefore, in the best of all possible worlds, and this conclusion, which so infuriated Voltaire, follows ineluctably from the very concept of God as a perfect being. He is not only the Creator of everything but also the sovereign and loving father of His human subjects. We might wonder whether this argument could convince someone who was dying of hunger or being tortured to death, but the question does nothing to shake Leibniz's faith; he does not deny that evil and suffering exist, only that their existence is a good argument against God's goodness. Theodicy, the theory of divine justice, was for Leibniz (who coined the term) a subject of constant reflection. Being is good, and it is good necessarily.

Leibniz's world suggests a number of questions, among them the following:

Leibniz says that if complex and divisible things (substances) exist, simple and indivisible things must also exist. Is this reasoning sound?

If we agree that God is absolutely good and omnipotent, can we reject Leibniz's claim, so often mocked, that we live in the best of all possible worlds? Should we not rather follow the Epicureans in saying that since there is so much evil and suffering in the world, God is either evil or powerless or both?

Faith:
Why should we believe?

BLAISE PASCAL
1623–62

Pascal did not want to be a philosopher. He did not think of himself as one, and it is probably fair to say that he wasn't one, at least not in the sense in which the term was used in his day. He had only scorn for philosophy and philosophers, and expected nothing useful to come from their disputes. Nor was he a theologian, in the sense that he did not engage in natural theology (which was actually a branch of philosophy, concerned with demonstrating theological truths through the use of reason). But no overview of European philosophy, however brief, could omit him. He died before he reached the age of forty. His most important work, known as the *Pensées* and published posthumously, might seem to be no more than a chaotic collection of notes and aphorisms – but they are aphorisms from

which European readers have been fishing out brilliant aperçus for almost three and a half centuries.

His interest lay in two quite distinct – though not, in his view, mutually contradictory – domains. One was mathematics and theoretical physics, the other faith and human destiny, both seen through faith and as it is without faith.

He was a first-rate mathematician and genuinely loved the craft – for that was what he thought mathematics was: a craft; the most splendid craft to have been created by the human mind, but a craft none the less, and of no use when it comes to the things that matter most in our lives. As a mathematician and a physicist he respected the modern rigorous rules of his craft; he had little use for the philosophical speculations of Aristotelians, with the aid of which semi-educated scholastics thought they could solve the problems of physics. He also decried appeals to religion in questions of reason or fact, and considered that the most flagrant example of such abuse in his own century was the condemnation of Galileo by the Catholic Church; do what you will, he said, the earth will continue to turn around the sun, and you with it.

But he considered the opposite manoeuvre – the attempt to demonstrate religious truths by arguments from observation – equally illegitimate and vain. Pascal was a man of unwavering faith; faith

was for him the most important thing in human life. But he knew that the alleged proofs of God's existence derived from natural phenomena were worthless; he knew there was no reliable way from birds, stars and the heavens to God the Creator. Nor can mathematical proofs help us in our suffering and misfortune; above all, they are powerless to teach us anything about our fate and its meaning, or about death. Human fate, he wrote, is like the fate of a gang of chained prisoners who must look on as their companions are put to death one by one, and know that they will be next, and that it will be soon; and when we realize that this is our fate, and that no voice comes to us from the unfathomable abysses of the universe, which is empty and silent, what are we to do? It is remarkable, Pascal says, that people do not think about the things that most vitally concern them: their deaths, immortality, salvation. They do not think about them because they do not want to think about them; they would rather not be reminded of what awaits them. They flee from what is most vital, escaping into amusements of all kinds, anything to forget; their entire life becomes a series of amusements, a way of escaping. We invent all sorts of ways to avoid confronting the fundamental issue: hunting, theatre, parties, intrigues, even wars – all these are just ways of anaesthetizing the pain of existence.

Faith can cure us. But how do we acquire it? Not from the silent universe, not from science, not from the observation of nature. Faith is not a series of propositions to which we assent intellectually, not even the proposition that God exists. Assenting to such a proposition might be acceptable and sufficient for a deist, who is almost as far removed from Christianity as an atheist. But faith is not assenting to propositions at all; it is plunging into a different reality. Faith is love of God and trust in God, the certainty of being one of God's children, the feeling of an unshakeable bond with the divine. These are not propositions demanding intellectual acts of assent; they are reasons of the heart. But though they flow from the heart and not from reason, they are not emotions or dim feelings that might be illusory. They are intuitions which bring certainty, although they cannot be proven in the strict sense. But the axioms of geometry also cannot be proven; they, too, are accepted by the 'reasons of the heart', just like our certainty that we are not dreaming when we are awake, despite Descartes' doubts.

But the question of how to acquire faith remains open. Pascal, like his co-religionists, the Jansenists (the Church condemned them as Jansenists but they themselves did not use this appellation, preferring to call themselves followers of St Augustine), believed that faith is a gift bestowed by God; it cannot be

acquired by natural means, by our own efforts or by scholarship. But God does not bestow faith as a reward for merit – for our good deeds, good conduct or orthodox beliefs. He bestows it upon whomever He chooses, gratuitously, arbitrarily, according to His pleasure, not as a reward but for reasons which are beyond our ken. Faith is the result of Grace, and Grace, by definition, is bestowed arbitrarily, not according to merit. Moreover, Grace is not only necessary for salvation but also sufficient for it, for when God confers it upon us, it is irresistible: we cannot reject Grace. Christian humanists found it hard to accept this image of God, a God who bestows salvation upon a select few according to His unpredictable whim, but this was the Jansenist doctrine and Pascal shared it. It is sacrilege, the Jansenists said, to accuse God of injustice, for God is the measure of justice: whatever He does is just by definition. Similarly, a sin is a sin because it is contrary to the will of God and for no other reason. Good and evil are good or evil not in themselves but by virtue of God's decrees. This flows as a necessary consequence from the fact of God's absolute sovereignty: no one, not even the most virtuous human being, can compel God to save him. God cannot be compelled to do anything, and our merits have no weight in divine judgement.

This terrifying Augustinian concept of divine

sovereignty and man's moral powerlessness is supposed to be a consequence of the theology of original sin, which radically altered our condition. The
post-lapsarian corruption of nature and the infinite
burden of Adam's sin, which weigh upon us all,
have, so to speak, stripped us morally: they have
divested our acts of moral power, so that nothing
we do, no act of our own free will, can lead us to
salvation. But is it not contrary to reason to believe
that God would condemn us to eternal torment for
something we cannot avoid? That He would consign even newborn infants to the fires of hell if
they die before they are baptized? Pascal acknowledges that the doctrine of original sin is unfathomable for us, that it goes against our reason and our
(miserable, inadequate) idea of justice; and yet, he
believed, without it our lives and our fate seem
more unfathomable still.

The Church's condemnation of Jansenism as a
heresy was, in effect, a condemnation of Augustinian theology, although the name of St Augustine
was not mentioned. It is a theology which may
seem horrifying, and that is indeed how it seemed
to many in the seventeenth century, the Jesuits
among them. There are passages in Pascal's *Pensées*
through which – despite Pascal's occasional assurances about the happiness of those who have been

blessed with divine Grace and have found God —
an element of desperation seems to creep. For if
faith is a gratuitous gift, it seems utterly pointless
for those upon whom it has been bestowed to try
to convert others. And yet Pascal makes such an
attempt, in the famous argument known as Pascal's
Wager.

The argument seems simple, but it has given rise
to misunderstandings, even outrage. It is a wager
about God's existence. Since reason cannot decide
the question, we must bet on it; and we must decide
which way to bet just as we would in a game of
chance, by calculating possible losses and gains.
The assumption is that we must bet, one way or the
other; we cannot declare that we will not play. This
is probably because it is assumed that if we do not
bet that God exists, we must be betting that He
does not, even if we do not say so explicitly, because
that is the easier choice in our earthly life. We must
now calculate the stake and the possible gains. The
stake, if I bet on God's existence, is my life here on
earth, which I must transform in the Christian spirit;
it is a poor, small thing, of little worth. But the
gain, if I bet on God's existence and win, is a life
of eternal happiness. Thus it would be in accordance
with reason to bet on God's existence, for if I win,
I win eternity, and the stake is small; it is worth it

to stake one life in order to gain three, let alone an eternity. In these conditions one would have to be mad to bet otherwise.

And if someone should object that he cannot believe, he must be persuaded that his resistance to faith does not spring from intellectual reasons but from the fact that he is enslaved by his passions; if he can subdue them, he will be removing the greatest obstacle on the road to faith. He must try, and trying means behaving like a Christian: going to Church and hoping for the gift of Grace. Pascal is not, of course, saying that by going to Church one can buy salvation, only that an obstacle will be removed. He surely means by this, though he does not explicitly say so, that by fulfilling the demands of the Church, even without faith, we will in the end transform our spirit, for there is a mysterious bond, the nature of which is unknown to us, that links the soul and the body. Moreover, if a possible convert changes his ways and abandons his sinful life, he will soon see that he has lost nothing, but gained much. Every would-be punter must surely have in his soul some will to believe, some vestige of hope, however weak. In any event, only God bestows faith, just as only He bestows salvation, so it is unclear how one can attain faith by one's own efforts, however tireless and sincere; God, if He so

chooses, can in one instant transform the heart of even the most hardened sinner: the Apostle Paul is an example.

It is impossible to prove the existence of God by reason. God is infinite and His ways are unfathomable; our reason must humble itself before His mystery. Even miracles and prophecies, though they are powerful signs of God's presence, will not convince unbelievers; no arguments will convince them, for their disbelief does not flow from reason, but from their uncurbed passions. God gave us all the illumination we need: unbelievers will not be able to plead ignorance at the Last Judgement. In religious matters there are all sorts of truths that we cannot grasp: Original Sin, the relation between body and soul, the creation of the world and the existence of God itself. We must recognize our own wretchedness, otherwise we will be guilty of the sin of pride; but recognizing one's own wretchedness without God breeds despair. Everything we know – both about our own wretchedness and about God – we know through Christ; He is our God, and without Him all human truths are powerless. Neither reason nor human history will save us. Perhaps the most famous of Pascal's aphorisms says: 'Cleopatra's nose: if it had been shorter, the whole face of the world would be different.' A world

without God is senseless chaos. But when we are nourished by faith, apparent absurdities disappear and everything becomes comprehensible.

People would prefer it, Pascal says, if Christianity were not true, because then they could happily go on indulging their passions, whereas the Christian faith tells them to disdain the body and worldly goods. Atheism is convenient. Scepticism is convenient. Scepticism is right in so far as it puts reason in its place and subdues its pride, but it is dangerous when it stands alone, without faith. But it is quite in accordance with reason to admit that there are many things which surpass our reason. Christianity is not a philosophy; it is not a metaphysical doctrine. It is life in faith.

Pascal suggests questions about everything. But we can pick out a few more specific questions:

Natural virtues are of no importance in salvation; an atheist can be a good person, but this will not help him at the Last Judgement. Does it follow that, since salvation is the only thing that really matters, it doesn't matter how we live?

If faith depends on the gift of Grace, can it be an illusion? Is it possible for us to believe that we have faith but be wrong?

Does Pascal's Wager not encourage the heretical hope that we can obtain Grace through our own

efforts? And was Pascal a Christian in the sense in which people think of themselves as Christians today?

Reason, Freedom and Equality:
What did God endow us with?

JOHN LOCKE
1632–1704

Although Locke's most famous work, his *Essay Concerning Human Understanding*, is immensely long, tediously repetitive and irritatingly vague in some of its most crucial points, it is deservedly considered one of the pillars on which Enlightenment thought reposed; his *Letter on Toleration* and his two *Treatises on Government* (especially the second) are also vital parts of the Enlightenment tradition. If some of his ideas seem like platitudes to us today, it is precisely because he made them so; they were not platitudes when he worked them out. He was not the first philosopher to argue that everything we know ultimately comes from sense-experience; the French sometimes claim priority for Pierre Gassendi in the matter of the primacy of experience, and they are probably right. But it can't

be helped that it was Locke's works, and not Gassendi's, that the world read.

Locke asks a traditional question: what are the limits of our understanding? What can we know for certain, what can be known only with a degree of probability, and what cannot be known at all? These limits, in his view, were narrow; he considered that most of the metaphysical questions which had preoccupied philosophers for centuries were either sterile or insoluble. He was a contemporary of Spinoza, but it would be hard to find two philosophers further apart in their thought than these two. As a student at Oxford he had to learn scholastic concepts and doctrines, but he soon came to the conclusion that there was little benefit to be derived from these studies and that they only bred confusion.

One might think, given his views on the narrow straits within which our knowledge must navigate, that Locke was a sceptic, but he was not: he was an ascetic. He did not lament or deplore the powerlessness of human reason when confronted with the metaphysical questions other philosophers had considered important and soluble. He simply thought that God had endowed us with intellectual faculties sufficient for all our practical tasks, and no more, but that there was nothing to be deplored in this; there is no benefit whatsoever to be derived from

addling our brains with theological riddles or puzzles about substantial forms and essences, for the only knowledge we really need for life is accessible to us.

The belief that there is some knowledge that is innate in us has been, says Locke, the source of much superstition and many dubious dogmas; anyone can defend whatever flight of fancy it pleases him to think up by claiming that it is some kind of innate wisdom, inscribed in his soul. The first chapters of Locke's main work are devoted to demonstrating that there are no innate ideas. The term 'idea' is a very broad one; it encompasses everything that can be the object of our thought, both qualities that are directly perceived by sight or touch, with their various combinations, and our thoughts about them, along with whatever we can discover by reflecting on the operation of our reason. But everything we know comes from experience; when we are born, our minds are like a clean slate, a blank piece of paper – a 'tabula rasa'. This famous claim of Locke's led some of his followers to conclude that all human beings are equal from birth, that none is better or worse.

The fact that we have no innate ideas does not mean, according to Locke, that our knowledge is limited to the contents of our perceptions and does not extend to things (as opposed to their qualities).

Sense-experience shows us only our own perceptual world, but we also have an unshaken intuition that those things exist, even though we cannot know their essences – but we have no need of such knowledge. This world of our perception is common to us all; it is not a collection of individual worlds. God endowed us with reason so that we could think about the objects of our perception, but reason, too, is the same for all of us. There are no grounds for thinking that sense qualities, which appear to us in various combinations, are all that there is; we must accept that behind them there is a substratum which is inaccessible to our experience and of which they are manifestations. Locke, like Descartes, distinguished secondary qualities (like colours and sounds), which are representations created in us by things, from primary or geometrical ones (like shape, size and motion), which are properties of things; but the mechanism whereby we perceive primary qualities in the ways we do is unknown to us. Nor do we know what material substance is, but, although it is not found in our experience, our supposition that it exists is well founded. Unlike Descartes, Locke did not believe that we can acquire knowledge of the soul or the soul's substance from the observation of our own mental activity.

We also do not, according to Locke, have any innate moral ideas; it is reason that teaches us what

Leszek Kołakowski

is good or bad, pleasant or unpleasant, beneficial or
not beneficial for us as societies. Locke was well
aware that such simple explanations could not be
a sufficient foundation for moral rules, but he
expected our reason to make further progress in
this domain, as it has in others.

Finally, unlike Descartes, Locke did not believe
that there was any innate idea of God; if there were,
we would all have one, but we clearly don't, so
there cannot be one. But we can be certain of God's
existence: for although we do not know His essence,
there can be no doubt that there must be some
entity which is the cause of our own existence, since
we, being contingent, cannot be our own cause.
This entity can also cause all the phenomena we
know from experience. We also know about God
from divine revelation, in the form of Holy Scripture. There are some things we can only know
through revelation, for they are not found in experience: resurrection, for instance, or the revolt and
fall of the angels. But we can justify the fundamental
truths of religion by reason, independently of revelation. Christianity does not conflict with reason,
and the truths of the New Testament are simple
and easy to grasp. However, if we find any claims
in the religious tradition which are contrary to
reason, we must not accept them. We believe in
revealed truth because we trust in God, but it is the

duty of our reason to examine each such claim carefully to make sure that it really is revealed truth. And here we cannot achieve perfect certainty, as we can in matters which are clearly and distinctly evident and where natural cognition can provide the answer. So we must be wary of sects which proclaim absurdities and call them truths which surpass the capacity of our reason to fathom them.

Thus, in general, Locke, although he acknowledged that revelation can sometimes provide answers obtainable from no other source, nevertheless insisted that no truth of revelation can be accepted if it is contrary to natural reason. The principle that we must prune religious tradition to fit the verdicts of secular reason is a deist principle, and in this sense Locke was a rationalist: he believed in universally binding rules of reason, the supreme judge of truth and falsity. He advocated the adoption of a general rule which is worth remembering: that we should believe nothing with greater conviction than is warranted by the evidence for it. This is a good and very useful principle, although it may not be universally applicable: we can imagine any number of practical undertakings over which a great degree of uncertainty hangs, but where success or failure will depend partly on our conviction of their success, even though that conviction may be contrary to the results of cold, objective calculation.

And there must surely be other examples of beliefs which we would do well to uphold with conviction even though the evidence for such conviction is lacking. One might also consider the case of self-fulfilling and self-destructive prophecies from this angle.

For Locke, the constant emphasis on the element of uncertainty in our knowledge claims, and also in our religious beliefs (though not the belief in the existence of God), is the theoretical basis of the idea of toleration. Since the Church is an association which we enter freely, and no institutions of the state have any authority in religious matters, we may not persecute people who think differently from us in such matters. Those who persecute others in the name of Christ are in fact betrayers of Christ; they cannot be servants of the truth. Thus the very nature of our cognition resists religious fanaticism and persecution. However, Locke did not go so far as to advocate extending toleration to confirmed atheists, for these, in his view, have no reason to keep promises or honour contracts.

Locke's analysis of cognition also turns out to be useful when we consider political arrangements and the conditions of a good political system. This is not the place to go into the conflicts within the British monarchy which led Locke to his theories about the state. His political theory inspired people

who were entirely uninvolved in those conflicts; it was an inspiration both to the French Encyclopedists and to the founders of American democracy. Locke thought that our reason is sufficient to establish what is just, and that God gave us freedom and equality in the state of nature. It is not entirely clear whether the state of nature is for Locke a theoretical fiction useful as an aid in reasoning or a state of affairs that really existed at some time in the past. But regardless of that, we know, Locke says, that freedom in the state of nature is our birthright: it is not innate, but we discover it as a gift from God. The right to life, liberty and property can also be respected in a state of nature, but the peace is often disturbed by those who infringe the rights of others; hence we need a social contract which will protect these natural rights and allow the government and the courts to restrain our desire to infringe the rights of others and to punish us if we do infringe them. When the distinction between what is mine and what is yours first appeared in the world, ways of deciding disputes about ownership were needed. There can be no liberty in a world where the right to property is not respected and regulated by laws. A social contract creates legislative organs and establishes a ruler who represents one side of that contract and accordingly takes upon himself certain obligations. It is not true, and could not possibly be

demonstrated to be true, that God made Adam the absolute ruler of the world and that contemporary monarchs are his descendants and inherit his privileges. A sovereign is the executor of the people's will, and absolute monarchy is contrary to the social contract. Only the people can decide whether the sovereign is honouring his contract with them, and when a king becomes a tyrant, citizens are no longer bound to obey him; they may revolt and depose him. In this way Locke's theory establishes the legitimacy of political revolution.

Liberty, property, political equality, religious toleration and the people as judge of the executive power – all these elements of the social contract are connected, and it was because of this connection that Locke became a classic of modern democracy. Again, if we think these principles are obvious, we must remember that they became obvious in large measure thanks to him.

Locke's writings suggest many questions, both about his theory of cognition and about his political theory.

The first question is this: for various Enlightenment thinkers, the idea that there are no innate ideas, that our minds are like blank pieces of paper when we are born, was proof that we are all equal, which of course in turn had important political consequences. But regardless of whether innate

ideas exist in the sense that Locke denied, we know that we are not born like blank pieces of paper, because we have genetically inherited traits which make us different in character and abilities. Should we conclude, in light of this knowledge, that the idea of equality has no foundation?

The second question is this: since the contents of our perceptions are the only source of our knowledge, on what grounds may we claim and consider as certain that objects exist independently of us, and that the contents of our perceptions represent them?

Finally, the third question: Descartes thought that the finite must be understood as a limitation of infinity, for the concept of infinity is intellectually prior. Locke, on the contrary, claimed that infinity was a negative concept, which appears when we are unable to determine or to imagine the end of a certain reality, for instance the end of a series of numbers or of space. Which of them is right, and does the question have any importance for us?

Perception and Causality: What can we know?

DAVID HUME
1711–76

When we talk about the cultural phenomenon of the Enlightenment or about Enlightenment philosophy, the word immediately brings to mind one of the chief slogans of that period: Reason. 'Reason' was the programmatic call of the Enlightenment, the basis of its legitimacy and its challenge to the age. Few people bothered about its precise definition; what mattered, and what was very clear, was the negative and polemical aspect of the word. Reason was the opposite of authority and tradition, of intellectual lethargy and the subjugation of the mind; it was a power which commanded us to scrutinize everything critically, to be fearless in questioning received wisdom, dogmas and superstitions. The famous Latin saying quoted by Kant, 'Dare to be wise', or 'Dare to use your reason', was

a clarion call that summed up the spirit of the age.

David Hume was perhaps the most brilliant, and the most elegant, exponent on Enlightenment thought in its strictly philosophical expression. He asked the familiar questions inherited from past philosophical tradition: what do we know for certain, and what do we merely think that we know? What are the limits of human knowledge? Where do superstitions come from? Can we trust what religion tells us, and if so, in what sense?

When Hume considers the reliability of our cognition, he starts from premises which are similar to Locke's, but more radical. The source of our knowledge is sense perception, or impressions, but also our emotions and passions: we see or hear something, love or desire something. Impressions are singular and immediate, but we make general judgements when we want to express certain resemblances among them. Only words can be general, not things or ideas or impressions: those are always particular. Apart from impressions we have ideas; in their simple form these are like pale copies of impressions, but combine and associate them in various ways; we can compare them and discover relations between them. These relations, which exist in our minds, boil down to three: resemblance, contiguity in space or time and causal dependence.

Impressions and ideas tell us about facts. But

apart from the knowledge about facts which we obtain in this way, we also know about certain relations between ideas from mathematics. In mathematics there are no facts; the truth that the three angles of a triangle are equal to two right angles would still be a truth even if there were no triangles. It is not a truth we know from experience; therefore it tells us nothing about how the world really is. It is an analytic statement, as Kant was shortly to call it: its truth can be determined by virtue of the meaning of its terms alone.

Apart from these truths, however, which tell us nothing about how things really are, what can we learn about the world from experience? Not much, it turns out. Yes, we have impressions, which in our experience are inextricably conjoined, like fire and heat, for example. But reason cannot demonstrate to us that this conjunction is constant and regular, that it always exists. In order to be able to say this we would have to be certain that nature is consistent and invariably operates according to immutable rules. But how could we be certain of this? Only from the fact that some of the things we experience, like fire and heat, always appear together; but to infer from this that nature is unvarying and consistent, and to go on from there to deduce that a permanent and unvarying connection

exists between these two experiences, or any other experiences, would be to argue in a vicious circle.

It is a circle in which we are invariably trapped whenever we talk about a necessary causal connection between phenomena. Of course, we have plenty of experiences where one thing is invariably followed by another: we know, for instance, that a stone, when we release it, will fall to earth and not fly up to the sky. But what is it, exactly, that we know? Only that in our experience it has always happened this way. We cannot know it *a priori*; we do not know that it must always happen that way. We know only individual events, and our belief that there are necessary connections between them results from habit and natural instinct, not from reason. The belief that such connections exist is essential to ordinary life and to our practical concerns: we would find life very hard if we did not expect stones to fall to the earth rather than fly up into the sky, or if we did not expect to be burnt when we put our hand into the fire. But we cannot know for certain that the future will be the same as the past; there is no certainty that the sun will rise tomorrow.

Our lives depend on our knowledge about causal connections, and yet it is a knowledge we do not and cannot have. Everything we know is hopelessly,

inevitably particular – individual impressions. We know no laws of nature, only our own habits; and we cannot deduce the truth from habits. Nor do we have any guarantees that our impressions reflect how things really are. There is no basis in experience for distinguishing between the primary and secondary properties of things, as Descartes and Locke do; since the qualities we see and hear are elements of our impressions of things, not properties of the things themselves, why should it be any different with such qualities as extension and shape?

We can see that the limits within which reason can manoeuvre are extremely narrow; it is not even certain what reason is for, apart from its purely negative tasks, like eliminating superstitions and the beliefs imposed by authority. Considering that we only know what individual impressions tell us, and that what we think of as knowledge of the laws of nature is simply the result of instinct or habit, reason appears to be a poor sort of instrument, woefully ineffective. And indeed, that is how Hume sees it, and sometimes laments its deplorable incompetence.

If this is the state of our cognition, we can imagine what Hume will have to say about religion and religious truths, and how he will dismiss them wholesale as mere superstitions.

But in fact Hume's verdict in matters of religion is, or at least seems, not quite so simple. He was

sometimes accused of atheism, and there were good reasons for the accusation. He prudently refrained from publishing his *Dialogues concerning Natural Religion* during his lifetime, so the work spent many dusty years in a drawer. It sometimes makes confusing reading, for it teems with ambiguities and oblique, allusive references – surely deliberate in a writer normally so scrupulous about clarity and precision. There are three interlocutors in the *Dialogues*, and it is not always clear which of them is expressing the author's own beliefs or in which passages those beliefs are expressed. There are arguments here (as well as in some of his other works) in favour of religious faith, but elsewhere in Hume's writings there are also plenty of arguments against it.

To say that God exists is to make a statement of fact; clearly, therefore, it is not a statement whose truth can be established *a priori*. But it is possible, or at least it seems possible, to argue for it on the basis of evidence that is derived from fact. One such argument starts from the ordered scheme of things in the world, and takes it as evidence of intelligent design: such a world must have been created by an intelligent being. But we know that we cannot argue from effects to supposed causes; such arguments can never be valid, for we can never demonstrate a causal connection. And we also know

that we will end up in a vicious circle if we try to demonstrate that nature is governed by immutable laws, i.e., that the world is ordered in accordance with certain rules. Moreover, the philosopher can hardly avoid noticing how imperfect is our world and its order; could God really be such an incompetent builder? Nor can we make any inferences about God's goodness or justice from observing those qualities in the world, for it would be difficult to describe the world as being absolutely and universally good and just. And if it were true that we acquire the idea of God by taking certain properties of the human mind and multiplying their power to infinity, the result of this procedure would tell us nothing about God's existence, only about the true source of our idea of Him.

Hume does sometimes speak of 'our holy religion'; he also says that it is based on faith and that attempts to ground it in reason are destructive. But the reader is at a loss to know where that faith comes from or what it is based on: on the gift of Grace? If so, how is Grace distributed? He sometimes talks respectfully about Holy Scripture, but we know that both the Old and the New Testament are full of descriptions of miracles and various kinds of divine intervention, and Hume argues emphatically that miracles do not exist. People's reports of miracles would be credible, he says, only if their

falsity could be considered a greater miracle than the event they describe. But this tends not to be the case. Not a single supposed miracle has ever been convincingly demonstrated as miraculous. It is significant, moreover (he adds), that most reports of miracles come from barbarian and unenlightened peoples.

As for the immortality of the soul, Hume has a separate essay about this which might strike one as startling. It contains arguments against the immortality of the soul: if something is indestructible and has no end, it surely cannot have a beginning either; is the soul eternal, then? And do animals have immortals souls as well, since they, too, feel and think? The connection between body and soul seems obvious to Hume: the organs of the soul belong to the body and, like the body, will decay after death. Our fear of death is in fact a good argument for the soul's mortality rather than the opposite: since Nature does nothing in vain, Hume argues, why should she have inculcated the fear of destruction in us if destruction were not really to be our fate? Since Nature is concerned only with our lives here on earth, she would be an astonishingly unjust Nature if there were another life that awaited us after this one. And can we really believe that God would be so cruel and unjust as to consign us to eternal punishment for the offences we commit

during our finite little lives here on earth? Suddenly, at the end of this long series of arguments against the immortality of the soul, there is one astonishing sentence in which Hume talks about the immense gratitude we owe to divine revelation, for only it can give us certainty about that great truth: immortality. The reader, having ploughed through the preceding arguments, can only conclude either that Hume is being very heavily ironic or that this is an equally obvious and heavy-handed attempt at last-minute prudence.

Hume's work was supposed to exalt Reason and Truth; instead, the picture that emerges is of the inadequacy of human reason and the poverty of our cognition. We cannot attain truth; we can know only individual impressions, which tell us nothing about reality; we extrapolate and go beyond these impressions for practical reasons, but when we do so we learn nothing about how things really are. In the end, reason turns out to be fairly futile; and Hume's occasional remarks about the natural order of the world, the aims of nature, God and revelation are quite contrary to the essentials of his philosophy. Thus poor reason, hardly born, committed suicide. Sad, in one so young.

Here are some of the questions Hume suggests:

Our whole lives depend on the belief in necessary causal connections, but there are no grounds for

such a belief. Individual impressions, on the other hand, tell us nothing about the world. It would appear that, according to Hume, the very idea of truth is a pointless fiction. Can we go on living comfortably if we really believe this?

We only know our individual impressions. But even in the simplest of phrases, we go beyond them. In the sentence 'I see a tree', for example, there is the word 'I', which may be – and indeed has been – questioned; there is the verb 'see', but is it legitimate? How do I know that I see, since I cannot see my own seeing? And so on. Can we imagine a language which contented itself with describing our individual impressions, without adding anything? What would it be like?

Do Hume's arguments against religion – most of them already familiar by then from other sources – appeal to the kind of general characteristics of nature, the world and reason to which one may not legitimately appeal within his philosophical framework?

Reason, Necessity and Morality: How is knowledge possible?

IMMANUEL KANT
1724–1804

The general question which preoccupies Kant will already be familiar, for it is the traditional question about the limits of human reason: what is human reason capable of and what can it legitimately claim to know? It has sometimes been objected that, just by asking such questions, Kant is entangling himself in a vicious circle, for in considering them we must make use of reason, and thus we are assuming from the start that reason is a reliable instrument. This is true, but the same objection can be made to all philosophers who ask about the limits and capacities of our cognition. There's no help for it: thought cannot start from zero, without any presuppositions. When we think, we are always starting from halfway down the road we want to travel, rather than

at its beginning; the very use of language makes this inevitable.

In order to deal with this question Kant divides statements into two kinds, which he calls 'analytic' and 'synthetic'. In analytic statements, the predicate is contained in the subject: 'All triangles have three sides', for example, or 'If John is the brother of Mary then Mary is the sister of John'. Here, if we know the meaning of the word which is the subject, we also know, without any additional information, that the predicate is true of it. Later philosophers described such statements somewhat differently: they spoke of propositions whose truth could be established by virtue of their meaning alone. Synthetic statements, by contrast, provide additional information, and are true or false by virtue of more than just the meaning of their words; their predicate is not contained in their subject. According to Hume and his successors (though Hume used different terminology), all synthetic statements are based on experience, and all our knowledge consists of just these two kinds of statements: analytic, or *a priori*, on the one hand, and empirical, or derived from experience, on the other. Consequently all that we know is either empty of content or inevitably uncertain, for sentences that describe our individual impressions do not allow us to deduce anything

about laws of nature or necessary causal connections.

Kant, however – and this is one of the fundamental tenets of his philosophy – came up with a third category of statements, which he called 'synthetic *a priori*'. These are statements that say something about the world but are also necessary and universally true. Such truths by definition cannot be derived from experience: we cannot deduce anything necessary from experience. But, according to Kant, if metaphysics is possible at all, if we can know anything at all about things that go beyond experience – about God, the universe, free will, the human soul and its immortality – then our knowledge of these things must consist of statements which are synthetic *a priori*.

This conclusion, however, appeared to deny the possibility of any rational metaphysics, for (it turns out) synthetic *a priori* statements can only be about the objects of possible experience. Our experience contains passive and contingent elements, but also necessary ones: space and time. Space and time are not autonomous entities, but constructs of our minds. We can grasp the idea of empty space, but we cannot conceive of physical objects outside space, without spatial characteristics; we can also grasp the idea of empty time, but we cannot conceive of events as taking place outside time. Time and space

are necessary forms, or categories, of our experience. They are not abstractions, but indispensable
tools of reason which the human mind has constructed. And they are among the things that synthetic *a priori* statements can be about. Geometry
and arithmetic, for example, consist of synthetic *a
priori* statements: they tell us something about the
world, but they are also necessary and universal.
The simplest proposition of arithmetic, '7+5=12',
is not analytic, for the predicate '12' is not contained
in the subject: 7+5 does not contain the idea of 12.
Similarly, we can prove that the three angles of a
triangle are equal to two right angles, but this truth
is not contained in the idea of a triangle. It turns
out, on Kant's analysis of synthetic *a priori* statements, that our knowledge – the knowledge that is
necessary and universal – can only be about the
objects of our perception.

In addition to space and time, there are two other
categories which the human mind constructs in
order to shape experience: cause and substance.
(Kant thought that the modern natural sciences and
Galileo's and Newton's physics, as well as geometry
and arithmetic, were all necessarily true in the same
way: because of these categories. Galileo knew that
physics does not describe experience; Newton's laws
cannot be derived from the content of our perception. The truths of physics are *a priori* truths.)

Physical objects as we know them are the way they are because we shape them by imposing these forms, or categories, upon them. This does not mean that the phenomenal world is an illusion or the mental construct of each separate human subject. Although the forms of phenomena – the categories of time, space, causality and substance – are the creation of the perceiving mind, they are a common, collective creation, so the world in which we live is common to us all. It is true, however, that objects cannot exist without us, the perceiving subjects. But the very idea of an appearance (which arises whenever we talk about perceiving, for appearances are what we perceive) presupposes that beyond the appearance there is something that appears – something else, beyond the appearance – although we cannot know what that something is. We cannot free ourselves from the subjective conditions of experience; we cannot know anything about how things are 'in themselves', except that they are, and that they are not the same as their appearances.

Some critics have objected that the concept of a 'thing in itself' is entirely unnecessary, for nothing changes if we eliminate it. But what was important for Kant about things in themselves was not just that their idea was inherent in the very concept of an appearance (for if something is an appearance, it must be an appearance of something), and thus that

there must be something – the thing in itself –
underneath it, otherwise we would have cause to
believe that the world was indeed just a figment of
our imagination. Particularly important for him was
the fact that our minds are constantly striving to
grasp those ideas which make sense of human existence and endow it with meaning: God, the soul as an
autonomous substance, the universe and its order,
which encompasses everything. Our minds cannot
help succumbing to the perpetual temptation to
reflect upon these things, even though nothing can
ever come of such reflection, for there is no way
from our experience to God, the soul or the universe. Kant considers each of the proofs of the
existence of God known to him and refutes them
all, one by one. He also points out that the fact that
we think does not entail the independent existence
of the thinking subject. Nor does our experience, in
which the principle of causal necessity is universally
applicable, provide any grounds for believing in
free will or in the immortality of the soul. This does
not mean that God and the soul do not exist, only
that we cannot demonstrate their existence from
facts or by the aid of the instruments of theoretical
reason.

But our reason does more than just interpret our
experience. Alongside speculative reason there is
practical reason, with its own *a priori* principles,

which are normative. These principles, or moral rules, are just as valid, just as absolute, as the *a priori* truths of mathematics. They concern our moral duties – and Kant assumes it is obvious that a duty is something that ought to be fulfilled just because it is a duty, never because it happens to agree with what we want or like or happen to feel like at the time. The supreme and most general moral rule, which Kant calls the Categorical Imperative, requires us always to act in such a way as we might wish everyone to act, and to follow only such rules as we could wish everyone to adopt. This absolute and universal requirement is dictated by reason; it shows that moral truths are the same for every rational being, and that our personal desires carry no weight in moral matters.

It turns out that practical reason has primacy over speculative reason and can arrive at metaphysical truths to which speculative reason has no access. Practical reason postulates free will, for although free will cannot be deduced in the world of experience, moral rules make sense only on condition that we have it. We must therefore believe that although we are subject to universal causal necessity just like everything else in the phenomenal world, of which we are a part, as things in themselves we are free. Similarly, practical reason postulates the immortality of the soul: this, too, is a condition of the

possibility of moral rules, because a moral injunction presupposes that the highest good is possible, and moral perfection is the aim of the human will; but moral perfection is possible only on condition that infinite progress is possible, and infinite progress, in turn, is possible only if our existence is infinite. Finally, practical reason postulates the existence of God, for the highest good (the possibility of which, we have seen, is entailed by moral rules) in turn entails the perfect harmony of moral perfection and happiness, two things which are not conjoined in the world of our experience; they can coexist in perfect unity only thanks to God, as two aspects of one world.

Thus it turns out that although the fundamental questions of metaphysics – the problems of God, of immortality and of free will – cannot be resolved theoretically, they find their resolution as postulates of practical reason: since reason demonstrates the validity of moral rules – for they are not arbitrary decisions, but are grounded in reason and flow from reason alone – and their validity in turn presupposes certain metaphysical truths, the validity of those truths is demonstrated.

Kant invented a peculiar philosophical vocabulary of his own to express some of these ideas, and this sometimes makes him difficult to read, but once we are used to this he is quite comprehensible. Here

are some of the questions a reading of his works suggests:

Various philosophers have claimed that the truths of mathematics are analytic. Let us take the apparently counter-intuitive statement that there are as many even numbers as there are positive numbers: is this statement really true solely by virtue of the meaning of its terms? Or the statement that the series of prime numbers is infinite: is it entailed by the very concept of a prime number? Or Fermat's famous theorem, which is very simple to understand but waited three hundred years for someone to come along with a proof: does it follow from the very concept of raising a number to a power? And if these are, as Kant would have it, synthetic *a priori* truths, what are they about? Time? Space?

The second question concerns the categorical imperative. Let us suppose that I ignore this imperative, so that, for instance, I would have everyone else tell the truth and keep their promises, but I myself intend to lie and break my promises whenever I feel like it. If I behave in this way, am I sinning against reason and falling into self-contradiction?

The third question is about practical reason. Is it possible that certain moral rules, not directly about God and the immortality of the soul, must neverthe-

less presuppose truths about God and immortality if they are to make sense?

The fourth question is this: is it true that our experience always, necessarily, contains a conceptual element?

History and the Absolute: Progress without good and evil?

GEORG WILHELM FRIEDRICH HEGEL
1770–1831

It would not be true to say that Hegel invented what we call the philosophy of history – which is the activity of reflecting upon the meaning and direction of human history – although in modern times he has been considered the driving force behind such reflections. St Augustine engaged in them before him, as did, in modern times, Bossuet. But Hegel's aim was different from theirs. Augustine and Bossuet wanted to reveal the hand of God behind historical events, to show how God directs the fate of nations and kingdoms, tribes and leaders, and whether He wants to enlighten us, warn us or punish us, and how He reveals to us the paths He wants us to follow in order to fulfil His plan for the world. God Himself, however, cannot, for them, be subject to historical change; being infinitely perfect

and immutable, He cannot be more complete or fulfilled than He already is, and thus cannot be guided by any aim involving His fulfilment or completion.

But a history of God is precisely what Hegel's work, viewed in this comparative context, can be said to be. Or, rather, it is a history of the absolute spirit. Here, the absolute is a historical reality (to the extent that the concept of a 'historical absolute' is not self-contradictory) which gradually reaches its fullness through the history of the world and of human culture, using these as the instruments of its own fulfilment. Such an idea is, of course, unacceptable from the point of view of any kind of orthodox Christianity, for it is unimaginable that God should practise His perfection and try to improve Himself.

The contrast is not, in fact, that sharp and clear-cut; it can be softened. For some Christian philosophers of a pantheist bent the idea of a historical God was not an absurd one (although they did not use the expression in their works). The great ninth-century theologian and metaphysician Johannes Scotus Eriugena, for example, wrote about the history of God: a God who, having created an imperfect world, tainted by contingency and evil, guides that world towards union with Him, and is Himself fulfilled only through that union. Yet the diversity of creation is not abolished at the end of this cosmic

drama, but endures in all its richness. God needs the act of creation: it is only through self-estrangement and a kind of self-negation, the negation of His own infinity, and finally by re-absorbing His creation anew, that He reaches His fullness. All the way up to the twentieth century, this idea, or a variant of it, can be found in the writings of Christian thinkers who succumbed to, or were touched by, the temptation of pantheism: dissatisfied with the traditional representation of God and the world, they sought an answer to the question of why God should have created the world if He is a self-sufficient absolute.

This is Hegel's question too, though differently expressed. His aim is to discover the full meaning of the history of the spirit. The spirit uses the history of human culture as an essential tool in its progress towards ultimate fulfilment; the history of mankind is thus part of the history of the spirit, which is both the beginning of the evolution of being (for Hegel this is a truth so glaringly obvious that it requires no discussion) and its end, with the achievement of perfect unity. In order to achieve such unity, the spirit must break down the barrier between itself and the object of its cognition; it must abolish the distance that separates them by assimilating the object entirely. For if the object is foreign to it, it is also contingent, and thus restricts the

spirit, which must encompass everything: it must *be* everything. This, too, is a problem inherited from Christian theology: God apprehends everything in Himself and through Himself; something outside Him, separate from Him, would be a restriction of His infinity. It is only when the spirit, through the travails of history, assimilates being in its entirety and becomes one with it, abolishing its contingency (but without destroying its diversity), that its inexorable progress towards its ordained fulfilment comes to an end, having attained its ultimate goal. This progress cannot go on infinitely, as it does in Kant's philosophy: if it did, the ground left to be covered would always remain the same – namely, infinitely long – and thus no real progress would take place. So the evolution of the spirit must have an end.

This evolution takes place through the constant negation of the previous stage attained. Successive negations abolish each previous form; sooner or later each form attained will be destroyed. But its destruction does not annihilate it completely, for its richness is preserved in the next stage.

We can understand this process because we partake of reason, which not only perceives the evolution of being but is itself also an element of that evolution. From this it follows that reason itself, as an instrument of the spirit, is changeable and relative. In other words, our perception of the world

and the things and events in it is always part of that world; it is not independent of it. This in turn leads us to suspect that until being attains its perfect, ultimate form, we cannot make any claims to truth in the ordinary sense of the word; at most we can make claims about historical legitimacy. This is a change of cognitive attitude: we are supposed to be aware of the fact that our thought is itself an element of what it thinks about, not an external observer or legislator – at least until being is entirely freed from contingency, consciousness unites with its object and the infinity of the spirit is fully realized.

Thus we cannot define reason through any of its supposedly eternally valid abstract rules. Yet we can be certain that the world, progressing ever higher in its evolution, is also becoming more and more rational: that its destiny is being fulfilled in its increase of reason. Hegel's most famous and most often cited dictum says that what is real is rational and what is rational is real. Some have construed this defiant view as the depressing injunction to acknowledge the rationality of every actually existing social or political form, however unacceptable when judged by our ordinary criteria, for even the most dreadful of forms merits our approval by virtue of the very fact of its existence. This would be a call to universal apathy. But such an interpretation

would not be quite accurate, for Hegel does say that we must first make sure that what we think we are observing at any given stage of evolution really is real, for its doom may be imminent: it may only be pretending to be real, but in fact will at any moment be swept away by the inexorable force of progress. Hegel neglects, however, to tell us how we are to go about making sure of this. We know that every existing stage must be annihilated, but we have no way of determining whether what we see is still historically well grounded or already gone, the memory of a dinner already eaten. In particular, we cannot make any assertions about this on the basis of abstract moral principles: by appealing to our idea of what should be the case but is not, to some idea of an ideal world. Hegel rejects such moralistic approaches to the world as pointless utopia-building: we are powerless to alter the inexorable course of events and it makes no sense to try. So in the end it does seem that applauding what exists, just because it exists, is the best, the only rational course of action.

This would be a poor result if it were Hegel's last word on the subject, the only thing to survive of his philosophy of history. But some of his other works contain a better explanation of the criteria of progress. The essence of the spirit is freedom, which

the spirit strives to attain by means of human history. Human history, consequently, is the progress of the consciousness of freedom. This progress – after Oriental tyrannies and European antiquity – achieves its expression in Germanic Christian culture, particularly through the Reformation, which, in Hegel's view, proclaims that man as such – every human being – is free.

This progress in human history is not the intended aim of the human agents through whom history happens; it is unrelated to their intentions. People fight for their private interests, are guided by their own passions and concerns, and yet they contribute to the realization of something unintended by anyone, but planned by the spirit, which, unbeknownst to them, is cunningly at work behind everything. History unfolds towards an end, but its end-directedness does not inhabit our minds. We are actors who do not really know the meaning of the play we are acting in, or where it will lead; we want power or riches or other earthly goods, but in striving for them we are unwittingly furthering the grand schemes of the spirit and leading it towards its fulfilment. When we know this, our approval of those schemes does not require a separate mental act of assent, but is entailed by, or contained in, the understanding; when we know it, we see a sense to human suffering and hardship. Individual acts of

rebellion against history in the name of moral ideals are pointless; they cannot change anything.

In the realization of the plans of the spirit the highest form of collective life is the state. It is in the state that historical destiny is embodied; it is also in the state that the reconciliation of the individual with the collectivity will take place. This reconciliation will not be imposed by force: it will be voluntary, since each individual will understand and freely accept his place and his tasks. For it is to the state that the individual owes the reality and value of his existence. And the state cannot make its decisions dependent on the will of the majority, for matters of state require knowledge and education (this is reminiscent of Rousseau, who distinguished between the 'general will' and the 'will of all'). We must also keep in mind that although we can comprehend political forms which are past, irrevocably gone and unchangeable, philosophy cannot anticipate or say anything about the future.

Hegel has been accused of totalitarian tendencies and viewed as a proponent of the omnipotent state. His writings have been seen as a collection of vague ramblings and bizarre ravings, sometimes self-contradictory, more often entirely incomprehensible, all swathed in a mantle of fuzzy phraseology. Nevertheless, regardless of the truth of this assessment, his influence on European culture was

tremendous, thanks in part, though not only, to Marx, who adopted – but also transformed – various peculiarities of his thought.

The questions Hegel raises are of a very general kind, and a full answer to them, if one is possible at all, would require a great deal of long and hard work. But they are questions we often encounter. Among them are the following:

What criteria can we use to measure progress in history if we refuse appeals to the idea of good and evil? (Setting aside the question of progress in technology and medicine though it is also interesting to ask whether there are any reasonable grounds for disputing the idea of progress in these domains as well.) And if such criteria exist, do we have any grounds for saying that human history is indeed a history of progress?

There have been occasions in the past, as we know, when the majority of a nation or a people enthusiastically supported criminal, destructive and murderous acts – crimes against humanity. In view of this, should we conclude that it is safer to abandon the idea of appealing to the will of the majority? And if so, on what basis should important decisions be made, and how can we make sure that they will not be murderous, criminal and destructive?

Philosophy, according to Hegel, cannot legitimately make any claims about the future; it can

only talk about past forms of social and political life. But let us assume that someone propounds rational arguments in support of the claim that even the most horrific forms from recent history, like Nazism and Stalinism, ultimately served progress; and that progress was similarly served by various earlier aggressive raids and invasions, because they brought historically higher European forms to primitive, undeveloped and stagnating lands. Given that appealing to the ideas of good and evil as criteria is not allowed, are such rational arguments possible? If so, could they change the way we think about these things? And if so, how?

World, Will and Sex:
Should we commit suicide?

ARTHUR SCHOPENHAUER
1788–1860

When we hear the name Schopenhauer, the image of a gloomy and bitter pessimist immediately comes to mind. Unlike the popular images of some other philosophers, that of Schopenhauer is bang on: a gloomy and bitter pessimist is exactly what he was. But that is only an assessment of his character, and Schopenhauer's pessimism went beyond character: he was also a pessimist as a thinker. And he was more pessimistic in his pessimism than any of the others who, along with him, created modern philosophy. He was the author of an enormously long work with the peculiar and disturbing title, *The World as Will and Representation*. The title is disturbing because while the thought that the world is a product of the human imagination, or a representation of something, is familiar from the works of

numerous earlier great philosophers, the phrase 'the world as will' seems, at first glance (and even at the second and third), odd. We can dimly discern what it might mean to say that the world is the object of our will, and the idea of the world as the work of divine will is even more familiar: a traditional idea, with nothing particularly scary about it. But 'the world as will'? What can that possibly mean? The word 'will', as we habitually use it, does not refer to any independent entity; we understand it to be an attribute or an activity of some entity, some subject, human or divine. Moreover, when we say things like, 'this is my will', or, 'that is God's will', we assume that this will contains some intention, some aim.

For Schopenhauer, however, 'the world as will' is not a metaphorical phrase or a spurt of extravagant language. He would have us believe that the world really *is* will. But not divine or human will; it is not the will of anyone or anything at all. Nor is it something with an intention, a direction, an aim or a plan. It is just a blind, aimless, impersonal, all-powerful force on which everything depends but which itself depends on nothing and no one; it just is. For Schopenhauer this truth – discoverable through acts of self-consciousness – seems utterly obvious, and he thinks it very odd that it should never have occurred to anyone before.

Schopenhauer adopted, but also transformed, the Kantian distinction between the phenomenal world and the unknowable thing in itself. There is no equivalent in his philosophy of practical reason, with its own *a priori* rules that allow us to discover truths about God, the immortal soul and free will. Here there is no God, no soul and no free will: everything in the world, and therefore also in life and in human behaviour, is governed by implacable necessity. In accordance with Kantian doctrine, the world of phenomena is co-created by the human subject, and time, space and causality are creations of the human mind. But the thing in itself, i.e., the true reality, the world that is independent of our mind, is will – unknowable, aimless and impersonal. Here is true metaphysical horror. We cannot subject this will to our control and we cannot know anything about it, and yet we must believe that it is will, and nothing else. Neither moral nor even mathematical categories are applicable to it; it is not evil or good; it contains neither number nor any other products of the human mind. Will also manifests itself in nature, both living and inanimate. But we are conscious of its operation, so there is a sense in which we can know it: we can know will directly through conscious inner experience, because in such experience the object of cognition and the cognizing subject are one.

Our most important experience of will is the will to survive, the will to life. The will to survive at all costs is present in every element in the world, and the will of each element inevitably comes into conflict with the will of every other. Thus the whole world of phenomena is a constant, endless struggle of all against all, as each will battles to survive at the cost of the rest of the world. This war, like the whole universe, manifests no meaning and has none. Everything that happens is the result of the operation of will, an impersonal and aimless force; there is no meaning in the world, and any meaning we think we find is an illusion. The human individual is an image of the universe in microcosm, and its life is just as meaningless as the life of the world.

All this we know, according to Schopenhauer, from experience. The metaphysical claims to *a priori* knowledge put forward by so many philosophers (very much including Hegel, whom Schopenhauer particularly loathed) are nonsense.

Can we nevertheless claim to have some access to truth? Schopenhauer certainly thinks that *he* does: his extraordinary quasi-metaphysical pronouncements about will as a thing in itself and its relation to the phenomenal world, and on the Kantian idea of objects as the constructs of our minds, are made with breathtaking self-assurance. He appears to believe he has discovered a great and irrefutable

truth about the world which, strangely, no one before him noticed. But at the same time he insists that our intellect is entirely the servant of our will, and cannot direct it. He also takes it upon himself to denounce the ploys and stratagems we resort to in order to see ourselves in a tolerably good light, and purports to reveal the self-deception to which we all notoriously succumb in order to hide, even from ourselves, what we know about our true motives and emotions (Freud says that Schopenhauer anticipated psychoanalysis here). Perhaps we are not capable of knowing the truth, after all – except, of course, Schopenhauer himself. Our consciousness is merely surface froth; we know nothing of the dark forces which really control us.

The picture of the world that emerges from all this is indeed a malevolent and lugubrious one: life is nothing but suffering and misfortune. Schopenhauer does not, for all that, advocate suicide as a solution. Not, however, for any of the reasons one might imagine, but because suicide, in his view, misses its target and does not solve the problem: a would-be suicide wants to kill the will to life, but by committing suicide all he kills is life; the will to life cannot be destroyed. Thus suicide cannot achieve its end. We ought rather to want the will to be able to operate without impediment. Who can make sense out of these arguments? They defy understanding.

Is there any other way of escape, apart from suicide, from this abyss of evil and suffering? Moments of respite, but only fleeting moments, can be achieved through aesthetic contemplation: the disinterested contemplation of works of art. For Schopenhauer the highest form of this contemplation is made possible by music: for music, unlike imitative arts, painting and sculpture, does not aim at objective representations of things, but wants to unveil the will, to reveal it and reach the nature of being itself; in doing so it frees us for a brief instant from the world of phenomena. It is unclear, however, why in fact we should derive any consolation from this, given how hideous and terrifying is the nature of being according to Schopenhauer.

There are, of course, moments of pleasure in life, but Schopenhauer stresses that pleasure is a purely negative phenomenon: it is merely the absence of pain. The pleasures of sex inevitably end in disillusionment. Schopenhauer even exalts the virtue of chastity (although he himself does not appear to have practised it much): the apparent satisfaction of our desires immediately gives rise to new, unsatisfied desires, he says. It is only in remembering the past and dreaming about the future – not in the present – that we can glimpse what is good in life. The world is bankrupt and its existence is no reason for rejoicing; it would be better if it did not exist.

The highest moral commandment is to exercise our ability to see other people as we see ourselves, to identify with others – in other words, empathy.

Thus there is no salvation, either temporal or eternal. So what is the point of life? This would surely seem a pointless question for Schopenhauer: the instinct for self-preservation, the will to survive, is part of our nature; it must therefore have been decreed by will, so it must be necessary and its destruction is impossible. The will wants what it wants because it wants it, and asking why will exists is a meaningless exercise.

Schopenhauer had a very high opinion of the classic Buddhist and Hindu texts, and thought his own image of the world was akin to the Buddhist one (however wildly implausible the idea that he created it under the influence of his reading of those texts). But Buddhism does offer a way to some sort of deliverance, whereas Schopenhauer's philosophy offers none. He did, however, adopt from the Buddhist tradition the principle that we should try to free ourselves from our desires instead of trying to satisfy them.

The questions that arise from a reading of Schopenhauer are odd ones, and they are about the fundamental meaning of his philosophy. Can we really understand the idea of an all-powerful and all-controlling, aimless, impersonal will, undirected

by reason? What kind of experience could make us believe in it? And how is it possible that no one saw the truth about it before, seeing that it is supposed to be evident through everyday experience? And why on earth should we, going against the ordinary meaning of words, call this will the source and driving energy of being?

The second question is this: what kind of experiences could we appeal to in support of the claim that pleasure and the good things we can experience in life are purely negative phenomena, the absence of pain, rather than vice versa?

Finally, a supremely metaphysical question: does it make any sense to say that it would be better if the world did not exist? Or is this an entirely meaningless thing to say? (Better for whom? Not for us, since we would not be there. Not for the cosmic will, because for it nothing can be better or worse. So just better in general? What could that mean?) And if the claim is absurd, is its negation – the claim that it is a good thing that the world exists – equally absurd?

God and Faith:
Do we need the Church?

SÖREN AABYE KIERKEGAARD
1813–55

It is impossible to summarize the content of Kierkegaard's writings. Among those who have the reputation of great philosophers but did not think of
themselves as philosophers, he, more than any
other, defies the attempt to describe his work: it
would be like trying to describe the work of a lyric
poet, which in a sense is exactly what he was.
The writings of no other philosopher are as tightly
bound up with his biography. Some of them are
barely veiled accounts of his private life – like the
third-person account of his love affair and engage-
ment, broken off by him for reasons explained in
an extremely involved passage of psychological
analysis, which itself is supposed to be of philosophical significance. Anyone who wants to paint a
full picture, even a rough one, of his thought, must

describe that self-destructive love affair. He must also describe Kierkegaard's relationship with his father, who thought of himself as an unredeemable sinner, damned for eternity (mostly because, as a young shepherd, weighed down by poverty and misfortune, he blasphemed against God), and raised his son in an atmosphere of extreme piety amounting to religious terror.

The popular view of Kierkegaard is as the nineteenth-century precursor of twentieth-century existentialism, or existential philosophy, although neither he himself nor any other well-known philosopher, with the exception of Sartre, ever applied the term 'existentialist' to himself; they did, however, talk of the philosophy of existence. Kierkegaard would recoil with horror and distaste at the thought that he was embroiled in any kind of 'ism', even that he was considered a philosopher at all. But he knew who would be his heir: that most repulsive, for him, of all figures – the university professor. The opacity of his work, often vague, tediously meandering and bristling with poetic repetitions, was clearly a deliberate stratagem; he wanted to make it impossible for academic philosophers, a breed he despised, to summarize it. His writings are more like oblique, mysterious allusions than lectures.

Kierkegaard's contempt for philosophy was

neither the result of a whim nor a secondary, minor concern; it was deeply rooted in his thinking, as it was in Pascal's, whose thought (though not his style of writing, which was aphoristic, limpid and succinct) was in many ways akin to Kierkegaard's. He thought philosophy worthless, for both of the two fundamental realities which above all else define our lives, namely 'I' and God, are inexpressible and indescribable – although we must try to describe them somehow, even if only elliptically, indirectly. It is important to notice that when we describe something as inexpressible, we are not talking about the attributes of that thing but about the conditions under which it can be known, and that is a concern of philosophy.

Kierkegaard has great difficulty in expressing what it is he wants to say. He repeats a number of times that each human person is a unique reality, irreducible to the abstractions that science – and Hegel – resort to in talking about people. But it is not enough to say that each of us is something unique and unrepeatable, for no two stones or leaves on a tree are identical either. The point is that every stone or leaf can be exhaustively described with the aid of abstract terms, but the human person cannot be so described – not exhaustively; the abstract categories do not exist that could capture the uniqueness of the human person. And since our

language consists of abstract expressions intended for describing a variety of objects, it is helpless when it comes to describing the unique and unrepeatable reality of the self – the 'I'. Human individuals cannot be added up to make a sum; it makes no sense to say that there are two individuals or more. An individual is not an example of a species. Hegel failed to grasp this, or did not want to. He reduced everything, including human beings, to what was universal – to general schemes and categories. Human uniqueness has no place in his system.

My uniqueness – because it is not reducible to anything else, because it is not part of the world, belongs to no species and is not classifiable under any class of concepts – must reject all the claims of rationalist thought. For rationalism seeks only those truths which are universal and acceptable for all; it looks for scientific methods and rules of thought that are universally binding. So it is useful for all sorts of practical tasks, but it cannot capture or describe the two ultimate, uniquely important realities: the human person and God.

For the human individual does not stand only in relation to itself. It is not part of the world, but nor is it an isolated atom, drifting in a void; it stands in a relation to God and eternity. God knows everything about each human existence, but He speaks only to the unique person that each of us is. Christ spoke

to each of His listeners separately, not to a crowd. Individual existence is a reality that stands before God in penance, awe and fear.

There are three forms, or stages, of life. In the first, the aesthetic stage, the individual concentrates on itself and seeks private fulfilment and pleasure. It does not do this by virtue of any act of its own free choice; on the contrary, this stage is a refusal to choose. But underneath the aesthetic pleasure lurks a hidden despair. In the second, ethical stage, the individual lives according to universal moral laws and freely submits to what is universal, universally important and universally understood. If there were no other stage beyond the ethical, Hegel would be right. But there is a third and final stage: the religious stage. Having reached the religious stage, the individual no longer appeals to moral laws, but only to God. And we can turn to God only in the awareness that we are sinners, in fear and in the recognition that when we stand before Him we are never right.

Kierkegaard considers the transition from the ethical to the religious stage in minute detail and from every possible angle, using as his example the terrifying story of Abraham and Isaac. By accepting God's command to kill his only son and thereby becoming a knight of the faith, Abraham makes it impossible for anyone to understand him.

Faith, for Kierkegaard, is not a series of dogmas or religious truths to which we assent or should assent. It is not a form of cognition. It is, as the Apostle Paul and Luther put it, unconditional trust in God and surrender to God. By the same token, faith is an acceptance of paradox and even absurdity. It is a paradox that God was able to assume human form; but He had to do so in order to transmit His teaching to us. The doctrine of Original Sin is a paradox: it is paradoxical that we should be held responsible for something we had no part in. It is absurd that Sarah should give birth to a son at her age, and it is absurd that Abraham should be commanded to kill that son. Abraham is either a knight of the faith or the murderer of his own child. In ordinary human categories he is a murderer. But in his case we must suspend universal ethical rules. His case teaches us that the human individual who is devoted to God stands above the ethical, above universal moral law – and indeed St Paul says that faith abolishes law. A knight of the faith is someone quite different from the hero of romantic or ancient tragedy; any resemblance is merely superficial. Iphigenia is also sacrificed by her father, Agamemnon, but this sacrifice is accomplished for different reasons: it is done in the name of social or tribal interest. Agamemnon is therefore obeying certain universal rules. Abraham's case is quite different;

he is not a tragic hero. And because of his faith his son is returned to him.

Faith is passion, a passion irreducible to the species of the individual. Faith is born when paradox meets reason. But it is not given once and for all; it must be continually renewed by each one of us separately. It cannot come from historical knowledge. Christianity is a historical fact, but it is not, *pace* Hegel, simply a stage in the evolution of the spirit. This is why history cannot teach us Christianity; we must be contemporary to Christ. To be an 'objective' Christian, through the institution of Christianity, is to be a pagan, Kierkegaard says. The institution of the Church as Kierkegaard knew it was, to him, the contrary of faith, a travesty of it. The Lutheran Church was not, for him, a place of faith or the repository of the Word of God. He saw pastors as state officials, bureaucrats busy with the advancement of their careers. It was to them that Jesus was referring when He condemned the Pharisees and chased the merchants and moneychangers from the temple. If He were to return to earth today, they would kill him. Kierkegaard's anti-Church diatribes sometimes recall the young Luther; and his accusation that Luther in the end founded his own Church recalls the accusations of the sixteenth-century radical reformists, who also believed that true servants of God have no need of any Church.

There are no rational grounds for faith, nor can there be such grounds. The existence of God cannot be demonstrated by rational proof, for there is no way of starting such a proof; if we try to construct one, we will inevitably be presupposing our conclusion in our premises. Once again Kierkegaard repeats: faith is not about theological assertions or arguments. It cannot be justified by argument and we should not attempt such justification. Nor is it merely emotion, as certain romantics claimed – although it is passion (but Kierkegaard never explains what passion is). God is that which is absolutely different and has no name; we are conscious of the infinite distance which separates us from Him. When reason tries to reflect upon something that is radically different, it falls into self-contradiction. Kierkegaard seems to be saying that faith is something entirely alien in the world of reason and temporal life. It is a gift from God, not from the Church – though Kierkegaard, in his attacks on the Lutheran Church, does not explicitly say that the Church is quite unnecessary and useless.

Kierkegaard is not a mystic, for he does not think that the infinite distance between the human person and God can be overcome; but he has in common with mystics his dismissal of attempts to provide a rational defence of faith and his belief that God is nameless – that He is, as Isaiah says, a hidden God.

Kierkegaard's view of the relation between the world and eternity emerges from his remarks about the sign of the Apocalypse. A fire breaks out in a theatre, but it so happens that it is Pierrot the clown who has to announce the fact to the audience. Of course, no one believes him; everyone thinks it is just a joke, and they go on laughing until they are consumed by the flames. That, says Kierkegaard, is how the end of the world will come: in the midst of general hilarity.

Kierkegaard did not want to be a philosopher or to be known as one, and yet, in wanting to express the inexpressible, he was one, in spite of himself. Accordingly, his writings suggest certain questions, although he does not express them in so many words.

If faith is, as Kierkegaard claims, a gift bestowed by God, not by mankind or by history, and if it is bestowed upon each of us uniquely and separately, ought we to conclude that religion can dispense with the Church?

Some philosophers think that the word 'I', although our use of it is legitimate, does not refer to anything. According to Kierkegaard, the unique and unrepeatable human person cannot be described in words. Are these two claims equivalent?

Can I say, meaningfully and without contradic-

tion, that something is true but only for me? Or that God is nameless and unknown but must be worshipped?

The Will to Power:
Is there good and evil?

FRIEDRICH NIETZSCHE
1844–1900

Nietzsche is sometimes called a nihilist, because his philosophy admits neither God nor any meaning in the world, nor any recognition of the Christian (or indeed any other) distinction between good and evil. He himself, however, far from describing his thought as nihilistic, reserved that adjective for condemning the Christian moral commandments, in his view hostile to life and to those instincts which can ennoble us. In the ordinary sense of the word it is Nietzsche rather than his Christian enemies who deserves the label. But this is merely a linguistic question, and of little importance.

The work of a philosopher may, of course, like that of a poet or a painter, be considered a part of his biography; and Nietzsche's biography has been studied by scores of historians, psychologists and

psychiatrists, all hoping to find the sources of his thought in it. But in the case of philosophers, such study is of a slightly different kind than in the case of poets or painters, where knowledge of the life can be not merely enriching but essential to our understanding of the work; a philosopher usually aspires to producing texts that are clear and comprehensible by themselves, and do not need to be explained by reference to events in his life, his illnesses, or the peculiarities of his character. If a philosopher's work cannot be understood without such references, it is surely not worth studying; the same cannot be said of the work of a poet or painter. Nietzsche's writings, then, can be studied as texts; we can ignore his ailments and the fact that he spent his last years in a lunatic asylum in the grip of mental illness. After all, his biography had no influence on those numerous writers and thinkers who succumbed to the charms of his rhetoric – for there is no question that Nietzsche was a master of German prose, though one can dispute whether the great role he played in the spiritual life of the twentieth century was beneficial or harmful.

Critics have pointed out that Nietzsche's various writings contain a number of ideas – about truth, and time, and art – which are mutually contradictory, although he pronounced upon each subject with spectacular self-assurance. But the target of his

Leszek Kołakowski

attacks is clear: European civilization. He was not a philosopher in the sense that he did not wrestle with the questions posed by Locke or Kant; his aim, mercilessly and unflaggingly pursued, was to show how weak, contemptible and degenerate was the European civilization of his day, to show up its illusions, its falseness and its self-deception, its inability to see the world as it is. How, then, should the world be seen? Nietzsche's answer is clear: neither the world as a whole nor human history has any meaning, any rational order or aim. There is only mindless chaos, with no Providence watching over it; it is directionless, tending towards no end. There is no other world; there is only this world, and everything else is an illusion.

To be sure, by Nietzsche's time it was not as if atheism was something new in European culture; it was not a shocking or unprecedented act of defiance. But Nietzsche's most famous sentence, Zarathustra's proclamation that 'God is dead', had a spectacular effect. For it did not mean simply that there is no God. It was a blow delivered to the heart of the bourgeois culture of Germany and Europe, and its aim was to demonstrate that the world of this culture, shaped by and drawing its cohesion and identity from Christian tradition, had ceased to exist, and was deluding itself if it thought otherwise.

Nietzsche wanted to force people to recognize that the world was empty.

Was this godless, aimless and orderless world conceived by science? No. In some of his writings Nietzsche extolled science, and he was fascinated by Darwinism, which in his day was still a cultural novelty. Though Darwinism did not inculcate in him a belief in progress, in the constant improvement of living organisms through the mechanisms of evolution, he did approve of the idea of natural selection and the survival of the fittest. The idea that weaker, inferior specimens are eliminated and only the noblest and best of the species survive appealed to him, and he thought such a law should operate also on the human species: weaker individuals should die while the stronger survived and helped the inferior ones to die. He despised Christianity because it embraces the opposite principle: it would have us preserve and protect the wretched, the meek and weak, in defiance of the laws of nature and the laws of life. He thought Christianity was simply hostile to life. Nietzsche also espoused the Darwinian belief that man is an animal. This, of course, is something people have always known, but in his philosophy it meant that man is only an animal and nothing more.

He also believed in the theory of eternal return

the Stoic doctrine that the universe has a finite
number of parts and therefore also a finite number
of possible arrangements, so that each must be
repeated, and will be repeated an infinite number of
times, although vast abysses of time may separate
each repetition. Thus everything, every detail of all
our lives, will be repeated exactly as it took place –
although of course we will have no memory of our
previous lives. This is different from the belief in
reincarnation in Eastern religions, for it contains no
idea of progress or degradation, merely a monot-
onous repetition of the same thing. Nietzsche took
this fantasy quite seriously; indeed, he looked on it
as his scientific hypothesis.

But on the whole, although he sometimes spoke
in praise of science (scholars of Nietzsche even talk
of his 'positivist' phase), Nietzsche did not expect
it to produce Truth in the full, dignified sense.
Science, like all cognition, cannot free itself from
its biased perspective. It is supposed to be based on
facts, but there are no 'facts', Nietzsche says, only
interpretations. And we do not need science to tell
us that heaven is empty and God is not there.

Can we find any meaning to our existence in the
Nietzschean chaos, any way to live in the belief that
life is worth living? There is no meaning that is
already given, waiting to be discovered; no external
meaning with which our life is endowed. But we

can create one. In order to do this we must reject popular superstitions and abandon the precepts of Christian morality, which exalts compassion, weakness and humility and is hostile to life. It is a morality born of resentment and the desire for vengeance – the morality of slaves, helpless herds who dream of taking revenge on their noble masters. We must replace the morality of the Gospels with a powerful affirmation of life, and arm ourselves with the will to power, the morality of the noble and strong: the masters. This morality has no need of the distinction between good and evil; that is a distinction created by the morality of slaves. The word 'evil' is unnecessary; only the word 'bad' is meaningful, and 'bad' means hostile to life, hostile to the expansion of the strong, triumphant life.

But the 'will to power' – a phrase for ever associated with Nietzsche – can be understood in a narrow and a wider sense. In the narrow sense it describes the proper spirit of the noble and brave warrior who aspires to rise above the weak and contemptible creatures around him and is not afraid to be isolated and hated by the crowd; he is the incarnation of a higher form of human being. If there is any aim towards which humanity strives, or any end to which it naturally tends, it is realized here, in these, its best sons – though what 'aim' or 'end' could mean in this context remains mysterious. But in its

wider sense, which emerges in Nietzsche's later writings, the will to power is a mechanism that operates in the universe; one might call it a metaphysical principle, were it not for the unsuitability of such an adjective for this philosophy. Reality is a collection of an infinite number of centres of will to power, each of which struggles to enlarge the domain of its power at the cost of the others. Each of us is such a centre. But there is no direction, no aim and no meaning in this struggle.

The higher form of human being, guided by the will to power, is not concerned with his own private interest; but he is also ignorant of compassion, conscience and sin. This does not mean that he is a sadist who wants to inflict suffering on inferior examples of the species; he is simply indifferent to them.

A violent contempt for ordinary people, people without great intellectual or artistic aspirations, emerges from this contrast between the morality of masters, who strive to achieve greatness, and the morality of slaves, wretched, despicable creatures condemned to destruction – the rabble, as Nietzsche calls them. And yet Nietzsche must have known that he would find life very hard if he could not go down to the baker's on one corner of the street or the shoemaker's on the other. But bakers and shoemakers were of no interest to him. His hatred

and contempt were directed chiefly at the educated 'rabble' – journalists, writers, politicians and philosophers, dealers in trendy ideas, propagators of the belief in the rights of the majority and in human equality – in a liberal or socialist version. He blamed them for the corruption and degeneration of European civilization and its inability to confront reality.

Bertrand Russell remarked that Nietzsche's philosophy can be summed up in the words of King Lear:

> I will do such things –
> What they are yet I know not, but they shall be
> The terrors of the earth.   (II.4.277)

His mockery seems not without foundation when we consider some of Nietzsche's more extravagant boasts – that he was the prophet of the coming era, for example. Yet in this he was, to a considerable extent, right: we sometimes speak of the twentieth century as the post-Nietzschean era. Moreover, in his message to the future, Nietzsche does to a great degree succeed in expressing the nihilism, cynicism and atheism of our world, as well as its abandonment of compassion, brotherly love and other traditional Christian virtues. His image was, of course, tainted by German National Socialism, which held him up as its herald. But it had to distort him and

pick selectively from his writings in order to do so, for Nietzsche was neither a Nazi nor an anti-Semite, and this is not difficult to demonstrate. Had he lived to see the Second World War, he would probably have approved of the imperial designs of the Third Reich, even if they involved the extermination of other nations, on the principle that they were in accordance with the laws of nature – the bigger fish swallows the smaller – and it would be absurd to waste one's tears on lamenting the laws of nature. But he would have had only contempt for the Nazi rabble, for these were not the lonely and noble (though compassionless) warriors of his philosophy but the embodiment of the herd instincts and emotions he despised. So the 'Nietzscheanism' of the Nazis was fake, though only half-fake.

But behind Nietzsche's noisy exaltation of life, his apotheosis of all that is lofty, powerful, noble and great, one can sense helpless flutterings of despair: the incurable despair of a mind wounded by the discovery of the meaninglessness of existence.

We might ask – and common sense provides the answer, although statistics cannot – which group was the larger: those who, inspired by Nietzsche's rhetoric, attained some kind of perfection in some domain of human endeavour, or those who, finding themselves unable to do so, committed suicide, in obedience to his philosophy?

Here are some of the questions Nietzsche suggests.

Let us suppose that we have been converted to the belief in eternal return: the belief that we will repeat our lives, in the smallest detail, an infinite number of times. Would this be a pleasant prospect or a terrifying one? Or an indifferent one?

Nietzsche claims that Christianity is a religion of weakness and fear, hostile to life and power. Can the fact that it triumphed and established its dominion over large swathes of the world be considered a refutation of this claim?

Nietzsche tells us to exercise the will to power and create the meaning of life for ourselves, regardless of traditional moral laws and inherited ideas of good and evil. How, on this view, does a great artist differ in his greatness from a great criminal? Are we to admire both equally, since both created the meaning they wanted in their lives?

Consciousness and Evolution:
What is the human spirit?

HENRI BERGSON
1859–1941

Creativity, initial spirit, the new, the nature of divinity, intuition – these are some of the ideas which bring to mind Bergson's metaphysics. He believed, contrary to the Positivists, that metaphysics is both possible and necessary to our lives; but he claimed, contrary to Kantians, that metaphysics can be constructed on the basis of experience – indeed, that it is worthless without such a foundation.

The starting point of his theory is the experience of time and motion. Time is the reality we experience most directly, but this does not mean that we can capture this experience intellectually. After all, the past no longer exists, and the future is not yet; the only reality is the present, and the present is real through our experience: it is a mental reality.

But the past and the future are each unreal in a different way. The future is absent in a radical sense: while anticipation is necessary both in science and in everyday life, when we anticipate something, rightly or wrongly, we do not thereby cause the future to be experienced as a reality; we do not bring it about. The future is not there; there are only certain regularities of experience. The past, on the other hand, is real: not real in the same sense as the present, but in the sense that it is preserved in our memory, and this allows it to partake of reality. Both memory and our experience of the present are mental phenomena; and time is a mental reality. What about the kind of time we know from equations in physics? That kind of time is not real time; it is a fiction, constructed on the model of space. But this artificially constructed time is indispensable to us, for without it there would be no physics, nor any other scientific knowledge.

Science slices the world up into fragments which can be manipulated. But the properties of things are abstractions, Bergson (a nominalist) claims; what is real is always unique in the world.

Do we, then, have no access to the truth about the world, a truth untainted by practical concerns? We do have such access, through intuition. Intuition unites with the thing in itself and grasps it from the inside, in its uniqueness. It is not clear what kinds

of things can be the objects of this intuition, but I am certainly one of them: through intuition I can gain insight into myself, my soul and the nature of my existence. But our experience of other people is another example; we can gain insight into other people through such intuition. The most perfect example is the mystic's contact with God, a very rare thing but enormously important, perhaps crucial, in the history of culture.

Our intellect assumes, instinctively, that the world is motionless: things are there and motion is added to them, as it were. We try to reconstruct motion starting from motionlessness, as when we produce the illusion of motion when we move from still to still on a reel of film. But in reality it is motion, not things, that is primary; things are not entities to which motion is added but mental crystallizations of motion.

It seems to follow from this that the world is essentially mental, or rather that it is mentally directed motion. That is indeed the case, says Bergson. Each of us is a physical body moving among other physical bodies and subject to all the laws of the physical universe. Our brains, too, are organs of our bodies and serve their needs. It does not follow from this, however, that our entire mental lives, our memory, are products of our brains. We know, of course, that various brain injuries

affect the memory, and this has led many people to conclude that memory, as well as all other mental phenomena, are functions of nervous tissue. Bergson devoted many years to the study of neurophysiology, and the picture that emerges, from a separate book and a number of his articles on the subject, is the following.

Our brain is an organ which selects what we perceive and remember, omitting whatever is unnecessary. Pure memory preserves everything, every detail of our past experience, but the brain's memory is censored: it retains only what is of practical use. Our memory is not physically engraved on our nervous tissue, only manipulated by it. Our nervous system is where consciousness meets body, but the two are separate entities. Consciousness is neither a part nor a function of the body, nor does it have any spatial properties. It is a continuous, indivisible and constantly increasing reality; it is time, not an object in time. Although its content changes with each individual experience, it is a continuous personal entity, always one and the same self. Consciousness cannot be a product of evolution, something that evolved as an instrument for the survival of individuals and species. Such an instrument would surely be an unnecessary addition to our other survival mechanisms.

Since consciousness – that is, memory – is neither

physically attached to the body nor created by it, although it is constrained by it, we must conclude that it survives death and the body's decomposition. Its immortality cannot, by definition, be deduced from any empirical data; it is a matter of religious revelation. The question of life after death is at the root of philosophical reflection; philosophy would be worthless if it could not discover the meaning of human existence and answer the question of where we come from and where we are going.

Human consciousness, then, the human mind and spirit, although there is a place where it comes together with matter and operates jointly with it, exists independently from matter; it can create new things which matter does not predict or anticipate, and it itself does not know them before it creates them, for it is free. The same may be said about the consciousness, or spirit, or mind, that directs the process of cosmic evolution and the evolution of life in the universe.

The very title of Bergson's most widely read work – *Creative Evolution* – already hints at his intention and the target of his attack, which was twofold. His argument was directed both against the popular image of Darwinism and against the belief in intelligent design – in an omniscient and infallible Providence which designed the world. The evolution of organic matter and species is a

creative process, driven by a great conscious energy – a divine energy; however, it is not preplanned and predetermined in detail, but takes place through a struggle of will with chaos, producing ever more perfect works of the mind and spirit. The idea that all evolution boils down to chance mutations, and subsequently a mechanical elimination of the worst and least fit examples of the species, is contrary not only to the essential nature of creativity but also to the very idea of evolution; such mechanisms operate only in relatively small areas, fragments, of evolution. On the other hand, the apparently contrary doctrine, according to which every stage of evolution is planned, down to its smallest details, by some great Architect, also destroys creativity and thus, like the first doctrine, destroys real time, which is reduced to being no more than a crank winding up the mechanism that produces the pre-prepared, oven-ready plan. The belief in universal and all-encompassing causality is an entirely unfounded dogma, a superstition of philosophers. But its negation – the belief in inexplicable cracks or gaps, mystery loopholes, in the evolutionary process – is equally pointless and sterile. Evolution is directed by a vital spirit, a force of life, which does not know in advance what it will create, but has a constant, built-in will to use the raw stuff of matter in the ever more perfect expression of the

spirit. This divine force is not by any means infallible; it operates by trial and error, often abandoning the creations it has begun, or even completed, in order to find a different solution to the challenges it sets itself. Inert matter is essential to it for the task of creation, but it is also an obstacle to be constantly overcome. The highest, perhaps the ultimate, product of evolution is the human species, qualitatively different from all others and equipped not only with instinct and intelligence – organs which sometimes conflict with each other – but also with the ability to create and the capacity for free choice.

The image of God that emerges from Bergson's work is not a detailed one. God is creativity and effort and love, and we know this from experience, not from abstract metaphysical speculation. He is not an immutable being. His existence is inextricably bound up with the stuff of His creation; creation from nothing is impossible. God does not exist within Himself, alone in His divine being; He is, as it were, the active spirit of the universe. Matter and creation stand in a negative relation to each other: God could not have set the process of life in motion without using matter, but the organisms and species which are created from matter strive to ensure their survival and to perpetuate their existence, and thus inevitably give rise to conflict. Nature is, as we

know, an arena of battle; and it seems we must conclude that even God Himself would have been unable to create a world in which living creatures did not battle to ensure their survival at the cost of others. Christian theologians, particularly Thomists, accused Bergson, with some reason, of the sin of pantheism.

Although our life, dominated as it is by practical concerns, inevitably distorts our perception and our thought, we can sometimes achieve a disinterested view of the world. This is what works of art do. Artistic expression is an attempt to unveil the uniqueness of things, instead of searching for resemblances and common features. The exception is the kind of art that is directed to the intellect, and reveals the contrast between human beings, who are capable of adapting their behaviour to the circumstances, and the mechanical or automatic; the aim of such art is to evoke laughter – as in life we laugh when a person behaves like an automaton. Bergson's entire philosophy shines through his treatise on laughter. But it also shines through almost every page of all his works, where intuition tries to break through the resistance of language like vital energy striving to subdue matter.

Bergson distinguishes two similar, though not identical, forms of life in the collective history of people and religions. Nature endowed human

communities with moral rules, customs and rituals, which individuals must obey and through which the community preserves its identity and cohesion. These rules are tribal rules: they are there to protect the community, and they differ according to circumstances. They protect the tribe, but do not encompass all of humanity; they ensure cohesion through hostility to other tribes. But there is another morality, not tribal but universal, encompassing all of humanity, that has struggled to emerge since ancient times, clearing a path for itself through the efforts of prophets and religious thinkers, but also great philosophers. It is this universal morality that tells us that people are morally equal and that tribal or national differences are irrelevant in our recognition of the dignity of each individual person. This morality does not evolve gradually from the morality of tribal societies but through a great religious leap, the leap towards an open society. Bergson believed in progress.

Similarly with the history of religion. Tribal religions are also products of nature. Their mythologies are there to protect the moral rules of the tribe and prevent our intelligence, which is easily swayed to selfishness and self-interest, from infringing on the collective interest and promoting the satisfaction of its own needs over those of the community. Mythologies and systems of magic are ben-

eficial inventions of nature; thanks to them our lives are not at the mercy of the haphazard operations of chance, for the world is peopled with purposively acting minds.

But alongside tribal mythologies a qualitatively different, dynamic religion arose, an active instrument of human spiritual perfectionment and a force which gives us access to the divine sources of being. Dynamic religion, the religion of humanity, operates through great prophets and mystics, although the Greek, Hindu and Buddhist mystics, however great their achievements, have never attained the power of the great Christian mystics, imitators of Christ. Their love for humanity is the love of God Himself. They cannot effect a sudden transformation of the world's whole spiritual life, but they leave their mark, and thanks to this there is slow progress in religious life too. Once again, Bergson believed in progress.

Bergson's philosophy suggests a variety of questions. He does not explicitly pose them and they have no answers, but they are interesting questions none the less. Here are a few of them:

If it is in God's power to create intelligent but disembodied beings, which is what we will be after death, why did He not do so from the start, instead of giving us bodies and exposing us to physical suffering?

Leszek Kołakowski

If analytical reason and science must, in order to be effective and useful, distort reality, and intuition has a hard job of penetrating the resistance of language, must the mystic also distort God in order to speak about Him in a meaningful way?

Are there any experiences that are both describable and absolutely certain? And if there are, is our empirical knowledge really built on such experiences?

The Foundations of Certainty:
What can we know and how can we know it?

EDMUND HUSSERL
1859–1938

It is hard to talk about Husserl without recourse to his own terminology, for his writings teem with neologisms. Nor can it be done without mentioning the sometimes radical transformations that took place in the course of his long life, a life of extra-ordinarily concerted effort and, more than any other, concentrated upon a single aim.

It was a Cartesian aim: to establish what can be known for certain, with a certainty that is utterly immune from doubt, and distinguish it from what we only think we know. And, having established this, to distinguish what is truly real, what truly (and necessarily) exists, from the constructs of our imagination and the things we accept on faith, because common sense tells us to.

Let us take the simplest of experiences: it is

midnight and I see the moon coming through the clouds. But does the thing I see really exist? And what is midnight, exactly?

Science can tell us exactly when and where the moon will appear in the sky; it can also measure the flow of time with great precision. But it cannot, according to Husserl, give us a true understanding of the world, for it does not consider – and is not interested in – what the things it predicts and measures really are. Science investigates many things, but not the act of cognition itself. It can prosper on the Humean principle that we have only individual impressions, in which we notice certain repeated successions of events and certain features that seem relatively permanent. On this basis we can establish scientific laws and classify things into categories. We do this for practical purposes: if we know the laws of gravity, we can calculate the path of a projectile or build a bridge. But are these laws real and true in the traditional sense of those words? Can we assert that they are necessarily what they are, and that the world is necessarily constructed the way it is? Can we claim that things are really the way we describe them? For instance, is the moon I am looking at now the same moon I saw a minute ago? Science, according to Husserl, just shrugs at such questions, which are of no use to it in its work.

Moreover, we reason – or ought to reason – according to the laws of logic. But what kinds of things are those laws? What kind of thing is the law of non-contradiction, which says that two statements which contradict each other cannot both be true? Is it an eternal and immutable law? Some have said that this is just how our brains are constructed, or how our psychology works, or our nervous system, and that we are simply unable to think any other way. We don't need to bother about whether or not the law of non-contradiction is *really* true, they said; the question is unanswerable. And possibly meaningless to boot. Saying that the law of non-contradiction is a moral principle (because without it we would not be able to communicate and collective life would be impossible), as the great Polish logician Jan Lukasiewicz suggested, does not shed much light on the problem either.

Husserl does not think that the validity of logical rules depends on the existence of the world. And he wants to know if those rules are really, necessarily true. With this end in mind he constructed a whole new branch of philosophy called phenomenology. The term already existed, but Husserl gave it a new meaning of his own. This new domain of knowledge about phenomena is a science with no presuppositions: it begins, as it were, from scratch, from zero. It does not accept any existing scientific

truths, neither empirical, arrived at by experiment, nor deductive; on the contrary, its aim is to endow all sciences, all areas of knowledge, with meaning, by showing what it is that their enquiries are really about, what the things they examine really are. The task of philosophy is to return to the traditional, Platonic idea of truth – the kind of truth that cannot be measured by its applications or its practical benefits. We want to know not only how the world is, but how it must be.

Thus philosophy is to be entirely autonomous, independent of any existing knowledge. In order to clarify the field of its enquiry, it must set aside all the things we think we know about the world, all allegedly certain natural truths, in particular the belief in the world's existence or non-existence. We will not be able to decide this question, but for the moment we will suspend it; we will put it, as it were, in parentheses – the famous Husserlian parentheses. This procedure, called phenomenological reduction, is supposed to open up before us an infinite realm in which we can investigate phenomena in their necessary relations without assuming anything as to their existence. The proper objects of such enquiry, however, are not particular, accidental phenomena, but the real essences of things. So we need another reduction, called eidetic: this allows us to direct our attention towards universals, general entities, in

order to examine their nature. Universals are not autonomous entities; they do not have the status of Platonic ideas. But nor are they abstractions, derived from things through generalizing the resemblances between them; they can be known directly, unmediated by abstractions.

The realm of universals is a separate realm of being. The essences of things which inhabit this realm are correlated with human consciousness, but they are not mere psychological constructs; and this consciousness is not the ordinary consciousness which is the object of psychological investigation, but transcendental consciousness, or the transcendental 'I' – the pure subject of cognition, from which all trace of the psychological self has been removed. Descartes tried to reach the ultimate source of certainty by starting with the 'I', whose existence he could not doubt, but it was the ordinary, psychological 'I', the self, and thus a part of the world. Husserl, however, insists that the world, including the psychological 'I', must be entirely (though only temporarily) removed from the realm of phenomenological enquiry. What we then have left is the pure subject of cognition, the transcendental 'I'. Phenomenological enquiry is called 'transcendental' because it is independent of the psychological subject and of ordinary experience; it sets aside everything transcendent, i.e., everything

that goes beyond the subject in the way that physical things, according to the common view, go beyond it.

Phenomenological enquiry can encompass any domain: we can ask about the real essence of a work of art, for example. We know what a work of art is in the ordinary sense of the word (or at least we used to think we knew, until recently), but this is not what concerns us; we are not looking for an abstract definition, nor for the kind of knowledge that a historian of art needs for his enquiries, but for the true essence of a work of art. Similarly, we can ask about the true essence of number, or of the state, or religion, or causality, or time, or God. We do not assume anything about the existence of these things; one need not believe in God in order to pursue phenomenological investigations into His necessary attributes – His nature.

Do these two types of phenomenological reduction give us a foolproof method for arriving at the essences of things? Husserl seems to think that they do. But the purified field of consciousness will not be enough; we also need a special kind of phenomenological intuition, or 'insight', which will reveal those essences to us. But how can we be certain that this insight is authentic and that every happy owner of it will reach the same conclusion? There is no way to ensure such certainty. Thus it turns out that

there is no phenomenological method that will take us where we want to go. What, in that case, is this whole phenomenology and transcendental consciousness business for? It seems to be quite useless. One sometimes feels, when reading Husserl, that he is giving us an idea for something destined eternally to remain just a programme, something that will never be applied. His attempt to reconstruct all knowledge ultimately fails.

The theory of transcendental consciousness has one other important consequence (though it is never made explicit in Husserl's earlier writings; it is unclear whether he saw it as logically entailed at the time or if the idea arose later, independently). It is that the objects which present themselves to our scrutiny in this purified field are not only correlated with our consciousness, but somehow dependent upon it; they are constituted by it, or derive their constitution from it. It is not entirely clear what this is supposed to mean; it does not mean that they are created *ex nihilo* in an act of divine creation, but nor does it imply merely an intention directed towards something that is already fully formed. In the operation of consciousness the distinction between the act of consciousness and its object is never abolished; thus the object is not merely part of our consciousness. But if it is given to consciousness in a way that is obvious and

certain, if, in other words, transcendental intuition can take place, it is only owing to the fact that there is nothing mediating in the contact between consciousness and its object; although they are not identical, they are inextricably bound up. More than this: in the acts of cognition of this phenomenologically reduced consciousness, all objects, the whole world, are constituted by the subject. The argument for this is that nothing can be an object of consciousness if we do not make it an object of consciousness, so the idea that we could be conscious of something that is not an object of consciousness — i.e., that we could think about something that is not an object of thought — is an absurd and self-contradictory idea. This argument, which seems tautological and unconvincing, was used much earlier as an argument for a different kind of idealism.

Thus Husserl arrived at radical idealism: the very idea of a world that is not existentially correlated with consciousness is incoherent. If we remove consciousness, we remove the world. Husserl says — but he says this, as far as I can tell, only once — that if the world were destroyed, the 'I' would continue to exist. Thus he endows the 'I' with the status of a metaphysical absolute. Only consciousness is self-sufficient and autonomous.

The question arises of why Husserl's phenomen-

ology exerted such influence in twentieth-century philosophy, given that it fails to provide the infallible method it had promised and that none of the eminent philosophers for whom Husserl was an inspiration or a teacher espoused either his idealism or his method of phenomenological reduction. The reasons seem to be as follows. Husserl defended the autonomy of philosophy; he thought that if its task were restricted to commenting on the results of scientific enquiry, it would be uninteresting, sterile and incapable of standing by itself. Secondly, he defended the traditional, Platonic idea of truth, which was untenable if one accepted Hume's or Ernst Mach's empiricism; true knowledge tells us how the world really is, not what we should accept temporarily as true for practical reasons. Empiricism cannot answer the question of how we can be certain that our impressions reflect things as they really are. Besides, there is no reason to suppose that sense experience is the only possible foundation of knowledge. Thirdly, his critique of nominalism (which denied the existence of universals) was important, for nominalism seemed to be a natural consequence of empiricism. Fourthly, viewed from the neutral phenomenological position, the world of moral values – to which, according to empiricism, the idea of truth was not applicable – became no less valid a phenomenon than the products of science or culture

which were the objects of phenomenological investigation; it acquired equal status and legitimacy. After the collapse of the many attempts by German philosophers to construct great speculative systems, the doctrines of empiricism and scientism consigned the concerns of traditional philosophy to the rubbish heap, and this was for many people a cultural catastrophe. Phenomenology seemed to rehabilitate a great tradition, to be a defence against scepticism and relativism.

Thus, although Husserl was at pains to distance himself from the image of philosophy as a worldview and strenuously denied that his own philosophy was one, his thought was nevertheless a valuable source of support for those who sought a means of defence against the dominant philosophical currents of the day, like empirio-criticism or neo-Kantianism.

Husserl forced us to confront an uncomfortable alternative: either we accept the restrictions of empiricism, turning away from the great philosophical tradition – the search for truth, meaning and the nature of being – and impoverishing European culture, or we must accept some form of transcendentalism, not necessarily Husserl's reduction and his idealism, but the belief that the human mind can have some insight into being and truth.

Here are some of the questions a reading of Husserl suggests:

Can philosophy, or any other field of enquiry, start from scratch, with no assumptions about anything? Does not language itself already presuppose certain things about the world?

Let us assume what Husserl denied: that the law of non-contradiction is just an accidental property of our language, or an accidental peculiarity of our mental make-up, or an arbitrary moral norm. Can we imagine a world of creatures endowed with reason who do not know it or do not apply it?

Since everything we think about is the object of our consciousness, the idea that the world could exist independently of our consciousness is an absurd one. Is this reasoning valid?